DAYS OF RIVERS PAST

DAYS
OF
RIVERS
PAST

R.S. HOOTON

RMB

RMB | Rocky Mountain Books Ltd.
rmbooks.com
@rmbooks
facebook.com/rmbooks

Cataloguing data available from Library and Archives Canada
ISBN 9781771602204 (paperback)
ISBN 9781771602211 (electronic)

All photographs are by Robert S. Hooton unless otherwise noted.

Design by Lin Oosterhoff

Printed and bound in China by 1010 Printing International Ltd.

Distributed in Canada by Heritage Group Distribution and in the U.S. by Publishers
Group West

For information on purchasing bulk quantities of this book, or to obtain media
excerpts or invite the author to speak at an event, please visit rmbooks.com and select
the "Contact Us" tab.

We acknowledge the financial support of the Government of Canada through the
Canada Book Fund and the Canada Council for the Arts, and of the province of British
Columbia through the British Columbia Arts Council and the Book Publishing
Tax Credit.

TABLE OF CONTENTS

FOREWORD

RECLAIMING PARADISE

The unraveling of British Columbia's wilderness over the past century has been shocking in its scope and few have watched the diminishment of the province's great natural heritage more closely than steelhead anglers.

Like the remarkable fish they pursue, steelheaders are by nature extreme. They drive to the ends of the worst roads, climb down cliffs, wade to the edge of their skills, or go beyond, raft gnarly rapids and drag their boats through dangerous log jams in pursuit of their prey.

Not many have been more passionate about the chase, or more finely tuned to the damage being done, than Bob Hooton. Over the past 50 years, first as a fisherman and later as a fisheries biologist, he waded, boated, swam and caught steelhead in B.C.'s greatest rivers. He was among the first fly fishermen to explore some of the more remote waters, but as he laments in his new book, he was mostly too late - arriving after the impact of the logging industry, commercial fisheries and increased access to sports anglers had tipped watersheds into periods of rapid and catastrophic decline.

In this story of a lifetime of rivers fished he describes the rapid pace of change and catalogues the losses. It is not a pretty story and it is impossible to read it without feeling deep anger and regret. Anger over the failure of successive governments to protect one of B.C.'s greatest natural resources - a fishery so spectacular that it had no equal in the world - and regret that we, the people, didn't do more to demand its protection.

Despite that failure, it's not too late.

This retrospective, written by an authority on steelhead and a master fisherman, reminds us of what we once had and points to ways to halt the continued degradation. By reminding us of the past he gives us a chance not to repeat the mistakes.

Bob writes that he hopes his book will inspire a new generation of steelheaders to take up the fight. But I hope it does more than that - I hope it stirs in people more broadly the idea that our goal should not be to just stop the bleeding, but that we as a society must pursue a return to the natural, rich ecosystems we once had in B.C. Surely we should be striving to make the world better, not just managing its continued decline.

We need to do that not just for the steelhead, but for ourselves, for our own spiritual health and self respect as a nation. It simply is not acceptable to say we inherited paradise and ruined it, so that's the end of the story. We owe it to our children to at least start the restoration process, to start the pendulum swinging back towards where it once was. And as long as we don't forget what we had, there is at least a chance we can do that - we can reclaim paradise.

In his book Bob records the benchmarks for some of the rivers he got to before the ruination started and through his research shows where that natural mark was on watersheds that were badly damaged by the time he arrived on the scene.

With benchmarks we have clear standards - all we need to find now is the political will to set them as our restoration goals.

Over the past century we have not had governments with the vision or moral strength needed to protect or restore our environment. Instead we have had a succession of administrations that yielded to the dominant economic forces of the day - be it resource companies or the commercial/sport fishing industries - allowing projects that damage the environment on the basis that such activities create jobs and grow the economy. At the same time government has allocated its financial resources without any sense of balance, spending billions of dollars on new bridges and hydro dams, but almost nothing to protect and restore watersheds.

We need to break that pattern. We need to establish that destroying watersheds through sloppy logging or mining practices, or over-fishing stocks either by accident or design, are practices that simply won't be tolerated anymore. And we need our governments to start spending billions of dollars, not mere millions, on fisheries programmes.

It is easy to say politicians should fix this, that they should draft firm environmental laws and provide adequate funding for environmental agencies. But the reality is they won't change unless voters force them to. The public needs to demand more from our governments, and needs to demand it now. If there is a fundamental shift in public values, politicians will follow, if there is not, they will stay locked into a pattern that is degrading our natural world while brining profits to a few.

Coincidentally, during the years that Bob Hooton was exploring Vancouver Island steelhead rivers as a young fisherman, I was doing the same on the island's trout lakes and sea-run cutthroat streams.

What I saw on those waters parallels exactly what Bob saw on the steelhead rivers. I hiked into remote lakes surrounded by towering old growth forests, built driftwood rafts, and caught big cutthroats that now can only be dreamed of. Once the logging roads opened access, clearcutting valleys and despoiled watersheds, the 5 and 6 pound trout soon disappeared. The same thing happened in the sea-run cutthroat streams when the spawning grounds became paved with sediment washed from logging slash, and the fishing pressure increased.

Protecting watersheds from resource industry damage and limiting the impact of fishing doesn't require a great, complex management plan. What it requires is simply the allocation of adequate environmental protection resources, and the moral strength to do the right thing despite cries of protest from powerful lobby groups.

Throughout his long and distinguished career, Bob never had trouble speaking up for the steelhead, even if it made him enemies. Getting people in authority to listen was the problem. Despite that he fought for steelhead doggedly then and he is doing so again in this important, compelling book. It should be mandatory reading for anyone in government who cares about the environment and wants to avoid making the mistakes of the past. It is an urgent cry for action and one we as a society have to heed if paradise is to be reclaimed.

The rivers of the past haven't stopped flowing and steelhead haven't stopped running into them, but this book makes it clear we have reached a dangerously low ebb. Many rivers are down to remnant stocks. In some steelhead stocks have been extirpated. But there still are wild steelhead out there spawning naturally in clean gravel beds - and in that there is great hope.

Restoration is possible, but Bob Hooton makes it clear if we want different results, we need to do things differently. Surely the time to start that change is now.

- *Mark Hume*

ACKNOWLEDGEMENTS

It is not possible to cover all the people who have contributed to a half century's worth of history and experiences I've attempted to describe herein. I'll try to remember those who were most influential, with apologies to any who feel they have been forgotten.

No one was more supportive of the angler in me than my father. If only he were still among us to see where that all went! From a career perspective, it was a childhood neighbour, Bob McMynn, who set the stage. He arranged my introduction to then Assistant Director of Fisheries for the provincial government, Ron Thomas, who took a leap of faith and hired me. Like my father, Bob and Ron are no longer with us to judge the outcome.

Among the mentors I consider most influential as my professional career unfolded, I hold Zeke Withler, Chris Bull, Jim Walker (RIP), Dave Narver, Hugh Sparrow, Gerry Taylor and, especially, Art Tautz in the highest regard. All of them held senior positions in the provincial fisheries or habitat protection program at the time. Ted Harding (RIP) and Craig Wightman set high standards for a newbie like me to emulate during my early days in service. Colleagues George Reid (RIP), Rick Axford, Lew Carswell, Gary Horncastle, Sean Hay and Maurice Lirette were tireless office mates and field companions who shared my passion for the resource and its future through my Vancouver Island years. Charlie Lyons, a tremendous source of knowledge on British Columbia's fisheries management history, captained the Nanaimo ship. Later on, in Smithers, Jim Yardley, Reid White, Colin Spence, Bill Chudyk, Mike Lough, Dionys de Leeuw (RIP), Dana Atagi, Mark Beere, Jeff Lough, Sig Hatlevik, George Schultze and Ron Tetreau (RIP) made up team Skeena. Mike Whately was another valued colleague and friend whose career overlapped with mine in Victoria, Smithers and Nanaimo.

Steelhead Society stalwarts Ted Harding Sr. (RIP), Cal Woods (RIP), Barry Thornton, Peter Broomhall, Bob Taylor (RIP), Lee Straight (RIP), Craig Orr, John Brockley (RIP), Jim Culp, Rob Brown, Bob Mckenzie, Bruce Gerhart (RIP), Mark Walsh, Rory Glennie and Art Lingren, to name a few, were all respected advisors whose collective wisdom was an invaluable asset to those of us endeavouring to address mutual concerns. More distantly, professional colleagues and steelhead aficionados Steve Pettit, a career-long steelhead guy with Idaho Fish and Game, and Bob Hooton, recently retired after a long career with the Oregon Department of Fish and Wildlife, did much to educate me on and off the water we shared.

Lastly there is my wife Lori and our three children Raylene, Carmen and Brock. Their recollections of our early years on the Englishman River are more about a bountiful garden, neighbours with tractors and horses and freshly caught coho on the BBQ than steelhead, but one of us steelheaders in the family was probably enough. For my three girls this walk down memory lane will tell them much they never knew. For Brock, I think it will serve to fill in some blanks on how his own passion has developed. I thank him for becoming the single greatest reason I undertook this book.

PREFACE

On a blustery winter's day, while waiting out the worst of the wind-driven rain before the obligatory morning dog walk, I decided to do a bit of reorganization of my home office. Sorting through memorabilia accumulated over decades, I turned up an envelope I'd all but forgotten. Inside was a picture sent to me by my long-time professional colleague and friend from Oregon, the other Bob Hooton, following a trip we made to Gold River in the early 1980s. I'd tucked it away together with a photocopied page from the Spring 1987 edition of *Trout*, the regular publication of Trout Unlimited. At the time, the last page of each issue was titled "Backcasts." This one had a short piece written by legendary Atlantic salmon angler Lee Wulff. It spoke to a summer's morning on the Serpentine River in Newfoundland in 1940.

> It was fishing to revel in. In a lifetime of fishing none I can remember was more perfect. It was wonderful to have lived then, been there, and so enjoyed something that will never come again.

That quote and the picture of a gleaming, ghost-like winter steelhead with nary a single scale blemished gave me pause to reflect on things seen and learned over 54 years of pursuing steelhead, 37 of which were spent in the employ of the agency responsible for their management. It struck me that there was some history deserving of record.

A few of my contemporaries in the fisheries management business could do likewise, I'm sure, but none has ever taken that step. Among the countless anglers I have come to know there are those who have been prolific in terms of their offerings on where, when and how to "get 'em." A few accounts of their time and river specific experiences are out there if one knows where to look. However, no one I know has had the broad spectrum of association with steelhead in British Columbia that a lifetime of fishing and a career in fisheries management afforded me. That leaves me feeling obligated to put some of it between covers before time gnaws away any more of my passion for doing it. Perhaps this blend of observations, stories and occasional sprinkling of management history on a few of the rivers that left the greatest impressions on me will be of value for those who follow.

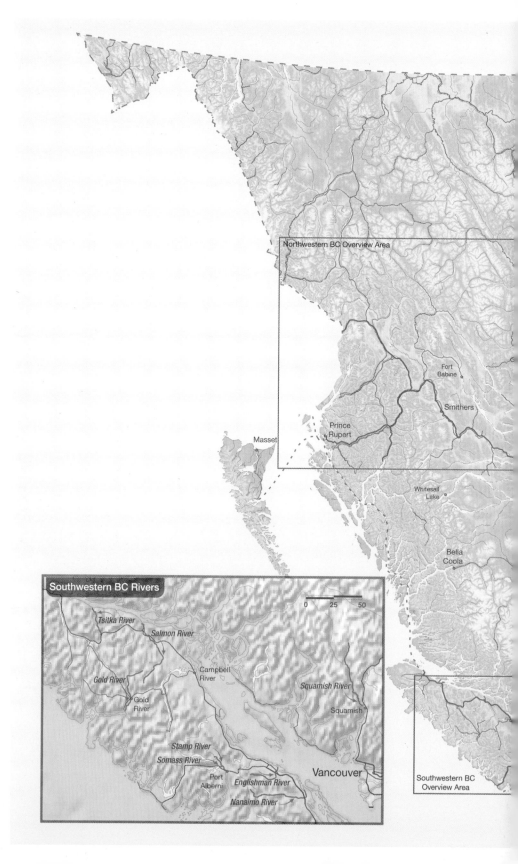

Northwestern BC Overview Area

Fort
Babine

Smithers

Masset

Prince
Rupert

Whitesail
Lake

Bella
Coola

Southwestern BC Rivers

0   25   50

Tsitka River

Salmon River

Gold River

Campbell
River

Squamish River

Gold
River

Squamish

Stamp River

Somass River

Port
Alberni

Englishman River

Vancouver

Nanaimo River

Southwestern BC
Overview Area

Northwestern BC Rivers

0 25 50

Skeena River

Sustut River

Babine River

Fort
Babine

Skeena River   Kitwanga

Smithers

Bulkley River

Prince
Rupert   Terrace

Fort
Nelson

Dawson
Creek

Tumbler
Ridge

nce
rge

Hixon

Quesnel

Mica Creek

Little
Fort   Golden

Clinton

Gold
Bridge   Kamloops

Kelowna

Cranbrook

Castlegar

Vancouver

Victoria

0   100   200
Scale 1:6,200,000

BRmB
backroadmapbooks.com
© Backroad Mapbooks

1

SETTING THE STAGE

A curious thing happens when fish stocks decline: People who aren't aware of the old levels accept the new ones as normal. Over generations, societies adjust their expectations downward to match prevailing conditions.

—Kennedy Warne[1]

Those words describe the evolution of fisheries in general the world over, but they capture the essence of British Columbia's recreational fishery for steelhead with disturbing accuracy. In compiling my recollections of times, places and circumstances relating to my favourite pastime, I was prompted by Warne's remark to review some of the more revealing history of what went before me. There are some compelling tidbits that speak to those times. In 1946, for example, Francis Whitehorse wrote:

> Looking to the future of freshwater sport fishing in the Province, it would not be less than truth to state that we have arrived at a parting of the ways: most of the lakes and rivers are slipping badly. Everything is against the fish: easy automobile travel; the combing of fishing water with power boats, and modern lures of fiendish ingenuity.[2]

Among those same pages Whitehorse cited records of his friend Arthur J. Milton, who lived on the north shore of Vancouver and fished the Capilano and Seymour rivers religiously. Over each of 16 successive years Milton never took fewer than 75 steelhead from those waters. His best year was 167. Milton passed in 1976, two days short of his 100th birthday. Oh what changes his fishing life bracketed!

More evidence of benchmarks of the past is readily available from the incomparable writings of Roderick Haig-Brown and his long-time neighbour and angling companion Van Gorman Egan (RIP). More on Haig-Brown later but, first, a reminder from Van Egan:

In the Nimpkish you can still feel the tug of a wild river, but an abundance of wild fish is not there. Nor is it in any other river on Vancouver Island. This island has been plundered and our expectations need to be tempered by half a century of bad logging practices and growth-at-any-cost attitudes. We struggle at great cost today to try to halt the perverse practices of yesteryears, to preserve what remains of a once abundant resource, whether it be forest or fish or samples of whole ecosystems. And it remains a struggle since the old attitudes are still with us, the greed and short sightedness and the threats by those who hold economic power to sway those who fear for their way of life, their jobs. It will probably be so, so long as there remains some economic richness coveted by the resource exploiters.[3]

Additional context for my own reminiscences stems from the journalism of Lee Straight, outdoors columnist for the *Vancouver Sun* newspaper from 1945 to 1978. I'm satisfied that his scrapbooks containing every one of his thousands of articles were never intended to be the continuum they represent today, but I view them as nothing less. Straight's descriptions of the fish and fishing that once existed in virtually every piece of flowing water between Squamish and Hope and so many other more distant places that emerged as frontiers during his career are invaluable history. How many present-day Lower Mainlanders would believe the Nicomekl and Serpentine rivers once produced regular steelhead sport or that a single pool on the Alouette could yield 22 steelhead in a day? How about six steelhead caught on a winter's day by a single angler working the bar below the Fraser's Pattullo Bridge? Cutthroat trout, once even more broadly distributed and more abundant than steelhead, are an equally distant memory among a rapidly fading angler demographic. There will be other references to Straight's material sprinkled throughout the pages that follow.

And what about the physical environment that supported the fish and fishing of yesteryear? In the era that fuelled much of the legacy of Haig-Brown and all of Straight's, the rivers of the Lower Mainland and those across Georgia Strait on Vancouver Island were unimaginably different than they are today. We baby boomers arrived too late to know personally what Haig-Brown and Straight had enjoyed, nor could we fully appreciate their frustration over the beginning of the erosion. Instead, we grew up during the most aggressive and unconstrained resource extraction era in the history of this province. Logging had a toehold long before, as did mining and commercial fishing, but dam construction, industrialization of estuaries and urban sprawl were just beginning.

Dams that supplied both water and power, mostly for the burgeoning Lower Mainland population, were the single greatest negative for steelhead and steelhead fishing. The Capilano (1954) and Seymour Falls (a project on the Shuswap Lake system, 1959–61) projects were obvious, but before them came the Coquitlam for both water (1904) and power (1914), and the Alouette diversion (1928) to feed powerhouses constructed at neighbouring Stave Falls (1911) and Ruskin (1930). Farther afield it was the Campbell, Quinsam, Heber and Salmon rivers that were all dammed or diverted between 1947 and 1958 on the strength of power demands trumpeted by government and delivered by the various forerunners of modern-day BC Hydro. The Puntledge River at Courtenay was added to the grid in 1955 and the Ash River near Port Alberni and the Cheakamus near Squamish in 1959. The Hydro engineers worked diligently to try to add even more rivers to their energy plans. Thankfully, proposals for the Fraser at Moran, the Skeena at Cutoff Mountain and numerous lesser-known waters met enough opposition from the likes of Haig-Brown to die on the drawing board.

Contemporaneously we saw the major forest companies develop sawmills of unprecedented size and log-eating capacity beside their pulp mills in or close to the best of Vancouver Island's estuarine habitats – the Somass at Port Alberni (1946), Nanaimo by 1950, Duncan Bay on the doorstep of the world famous tyee capital, Campbell River (1952), and then Crofton at the mouth of the Chemainus and right on the highway to the Cowichan (1957). Those were preceded by the dirtiest of all, the sulphite mills at Port Alice and Woodfibre adjacent to the estuaries of the Marble and Squamish rivers respectively. Their history dates back to the First World War period. Finally came Gold River in 1965. We could add another five mills on the upper Fraser and others at Prince Rupert and Kitimat, but those were distant from the waters I'll speak to here. All of those mills required power and water which, of course, came from the nearest river, often at the expense of late summer flows critical for fish. Most also consumed massive acreages for storage of logs to supply raw products for pulp and lumber. Those logs were commonly boomed and sorted over prime fish food–producing habitats for juvenile salmon and steelhead. Half century–old layers of bark and debris blanketing underlying substrates do fish no favours. In those early years quality control standards for mill waste discharges didn't exist, and there was no one to enforce them for years after they finally did. Thankfully, Houston on the Morice River and Ashcroft on the Thompson narrowly escaped the same fate.

Loggers built roads from gravel mined from river bars or taken directly from stream channels. Logs were hauled across and through streams whenever expedient. Stream hydrology had yet to enter the lexicon of the industry. If the forestry-related impacts weren't enough there was the mining effluent from Britannia that poisoned adjacent Howe Sound, the Mount Washington operation that did likewise to the Tsolum, the copper mine at the head of Buttle Lake, another near the Marble River and all the coal mining that affected so many rivers between the Quinsam and the Nanaimo. We mustn't forget, either, that gas pipeline and, later, the fast track highway to the southern interior that ravaged the Coquihalla River at Hope.

The population of greater Vancouver had reached roughly half a million by the time I was born. Today it is five times that number. The fields of Surrey, Langley and Cloverdale where I shot my first ducks and pheasants are continuous subdivisions, malls and industrial parks, save for off-leash dog parks, obligatory children's playgrounds and golf courses. I measure all those sorts of developments in steelhead generations. Fifty or 60 years, much less than one human lifetime and perhaps a dozen for steelhead, is all it took to transform the landscape and rivers irreversibly.

And what about steelhead and steelhead fishing? What direction were they heading in while British Columbia was undergoing all this progress? The retrospective glance reveals that there were some prominent myths afoot that were at the root of the trends that followed. Against that, one must understand there were no government staff employed in the fisheries management business until 1948, when Dr. Peter Larkin (RIP) was installed in the newly established position of Chief Fisheries Biologist. He resided in Vancouver, not the provincial government centre in Victoria. In fact there was no freshwater sport fishery management presence in the province's capital until 1962, and even then Dr. Larkin's skeleton crew of successors were little more than a voice in the wilderness. Details of all that followed with respect to the slowly developing rear guard of fisheries management staff responsible for steelhead are too much for these pages, but some underlying features are worth emphasis.

First, there were no steelhead licences until 1965 and therefore no reliable record of how many anglers were out there or what they were catching. Second, it was universally accepted that angling harvest was not responsible for perceived steady declines in steelhead abundance. Biologists and anglers alike trumpeted this message repeatedly from the first days of Dr. Larkin through to at least the early 1970s. By extension of that firmly entrenched belief, there was no need to reduce daily limits below three, possession limits below nine or seasonal limits below 40. In fact a seasonal limit was held to be meaningless in terms of any detectable influence on overall harvest of steelhead. Third, the worst of the alteration and outright destruction of steelhead habitat pursuant to logging was thought to be over or nearly so. Recovery of watersheds through natural regeneration of forest cover within 20–30 years was expected. Finally, and from an entertainment perspective if nothing else, there was endless debate about whether or not to ban the use of fish eggs or roe for bait. The arguments presented by the equally passionate opposing factions in the 1940s are indistinguishable from those prevalent whenever that issue surfaces today.

Fishing was a popular and prominent activity all through the 1950s and 1960s. Those years saw the steelhead angler population increase at an even greater rate than the general population. For British Columbia I don't think it inappropriate to refer to those times as the golden age of fish and game clubs. There were almost 200 clubs registered with the fledgling BC Federation of Fish and Game Clubs in the mid-1960s. Today there are only half that number. Numerous Vancouver-based groups dedicated to pursuit of steelhead were born and became top predators on most of the best waters at all the right times. They held regular derbies on the Vedder, Vancouver's north shore streams, the Squamish system, the Thompson and even on some Vancouver Island streams. The Drifters, the Streamers, the Squaretailers, the Black Bear Anglers, the Kingfishers and more all wore their colours loud and proud. Long-established fly fishing fraternities, notably the Totems and the Ospreys, also emerged during those years, but they were a softer presence and greatly outnumbered by non-fly fishers.

Fishing equipment was evolving rapidly at the same time the number of anglers was mushrooming. Fibreglass rods were replacing split cane, with graphite soon to follow. Centre pin reels were popular from the beginning and held their place among the aficionados, but new age spinning or threadline reels were popular among newcomers. As competition increased some turned to high-retrieve ratio level wind reels for competitive advantage. Fly fishing remained far more traditional far longer. The catch-up in recent years has been remarkable, though. Ever more effective and efficient fishing equipment made each generation of anglers a greater harvesting force than ever before. That trend has never changed.

Guides warrant mention here also. Prior to the early 1970s anyone who wanted to become a freshwater fishing guide need only to convince the local game warden that he was a reputable character to be issued what was known as a small game and angling guide licence. The fee was nominal, although

I've never been able to find conclusive evidence of precisely what it was. A prospective guide didn't need to be Canadian, didn't need to specify the exact waters to be guided, didn't need to demonstrate familiarity with regulations and wasn't required to file any report before obtaining the next year's licence. The first of the jet boats didn't arrive until the mid-1960s either, so no one had to be concerned about competition in that regard. No doubt there was the occasional guide operating on the Lower Mainland during my early years, but I never encountered one. On Vancouver Island there were only two guides, Len Francis and Gary Christensen, who pursued the business as a primary occupation for at least part of the year. Both were respected and highly skilled graduates of the Lower Mainland school of steelhead fishing. Francis set up in Courtenay in the late 1950s and became a legitimate pioneer of the steelhead fishing on numerous streams on the northern half of the Island. Christensen became his partner a few years later.

A few other features of the steelhead fishing scenario of the day need to be remembered. One was the persistence of commercial gillnet fisheries that targeted steelhead in the lower Fraser River right through the dead of winter. That fishery had existed for decades before there was ever anyone to examine catch numbers or what influence they may have had on conservation or the expectations of a steadily growing number of anglers. It wasn't until the 1960s that gillnets in the December through March period were forbidden. First Nations gillnetting and marketing of steelhead was another sore point among anglers. Foremost of concern was the Cowichan River. The difference between then and now is that the federal authorities of the day were quick to respond to complaints and there were consequences for anyone apprehended. Lastly, there was the issue of road access. Whereas the Greater Vancouver area was relatively easily accessible all through the post–Second World War years, the same cannot be said for other areas viewed as the grass beyond the mountains as the industrial onslaught unfolded. Logging companies had exclusive access to vast areas of real estate until the early 1960s. Vancouver Island exhibited the most glaring examples. It took years of relentless pressure from the first organized and politically astute group of hunters and fishermen of the day, the BC Federation of Fish and Game Clubs, to finally get government to force the companies to unlock their gates during "non-industrial hours" (nights and weekends). Until then many valleys laced with logging roads had been the private playgrounds of company brass, their families and friends and a few lesser-ranking employees.

Such was the river fishing landscape of southwestern British Columbia in the years leading to my love affair with steelhead. I knew almost nothing of the history of the rivers and the sport at the time, though, so I often reminisce about what it was that drew me into the game. I don't recall any particular event or occasion that set my course, but that boyhood neighbour mentioned earlier had much to do with it.

Growing up in South Burnaby in the 1950s and early 1960s our family home was three doors down the street from R.G. (Bob) McMynn, who was the Chief Fisheries Biologist for the province at the time. His son Robby and I were good friends. Tales of what his father did for a living would seem to have inspired thoughts on what my own future might hold. My own father had told me for years I shouldn't be thinking about following in his footsteps as a senior aircraft maintenance inspector for a succession of the country's major airlines. The fish business, as I saw it through my friend's father, looked and sounded much more attractive. A career would never have happened, though, in the absence of that meeting kindly arranged by neighbour McMynn with one of his Victoria colleagues at the time, Assistant Chief of Fisheries Ron Thomas.

I remember that session with Ron as though it was yesterday. We were mutually convinced, with me being a recent graduate of college as well as an apprentice in the Lower Mainland school of steelhead fishing and him a wizened fisheries manager, that steelhead populations in the rivers of southwestern BC were too large and robust for anglers to dent. Years later, with time in the professional ranks behind me, that perception would be far different. By then there wasn't a steelhead stream of any consequence on Vancouver Island or a good part of the mainland coast opposite that I hadn't at least flown over if not tramped along with electrofishing gear or a rod or snorkelled when we developed that as the learning tool of choice for assessing numbers, habits and habitats. I'd visited those islands once known as the Queen Charlottes often enough to assess and fish all their good steelhead streams. On the northwest mainland I spent time on every significant steelhead-producing tributary in the Skeena, Nass, Stikine and Taku watersheds as well as dozens of lesser-known outer coast winter steelhead streams. I even managed to see some of Kamchatka's magnificent rivers, albeit only for a couple of days of touring by giant Sikorsky helicopter. Those experiences put my steelhead world into a perspective that fit the times.

My good fortune in living near Bob McMynn and meeting Ron Thomas may have been instrumental from a career launch perspective, but there was one more event I now look back on as a key influence on so much that followed. It came during my first year as a bona fide fisheries biologist residing in the Nanaimo office of the provincial government's Department of Recreation and Conservation, the forerunner of several different ministries that housed the province's fish and wildlife management staff. In the midsummer of that year (1975) I managed to arrange a meeting with Roderick Haig-Brown. While still a student with a passion for angling and dreaming of a career in the fisheries business, I'd familiarized myself with Haig-Brown and most of his books. I firmly believed there was no one alive better able to enunciate a philosophy on resource management.

It was a bit intimidating to knock on Haig-Brown's door and introduce myself, but I was soon at ease with him in his study. The book collection was as impressive as the man himself. Our conversation was mostly about early days and memorable experiences on some of his favourite waters. Fly fishing for summer steelhead on the Heber River was a focal point. I asked if he still spent time there. In a reflective tone, without a hint of bitterness, he explained he couldn't bring himself to do that anymore. For one, the fish just weren't there in the same numbers. But, more importantly, the quality of the fishing he had once enjoyed was gone. The river had been diverted in its headwaters, a project he had rallied the public against and fought with vigour, and its valley and tributaries logged mercilessly. There was no going back. He went on to admit he had been spoiled by Atlantic salmon fishing in Iceland. Little more than one year later this giant among giants of angling literature, philosophy and environmental horse sense left us. I'm thankful I had that brief time in 1975 to be able to meet the person whose remarkably gifted, penetrating writings and countless public appearances and pronouncements on conservation issues have eclipsed all others through the history of natural resource management in British Columbia and beyond.

Today I can look back and understand fully that my tenure as a fisheries professional bracketed the period when such a career was as good as it ever got in British Columbia. The first 20 years were punctuated by growth in the environmental awareness of the public and the comparatively favourable response of the alternative provincial government party elected in 1972. That and a major federal government program intended to rebuild impoverished commercial fisheries (the Salmonid Enhancement Program, or SEP) injected new life into both federal and provincial fisheries agencies. By the late 1970s budgets were finally sufficient to have reasonable numbers of professional staff out on the waters building a long overdue knowledge base. My colleagues were well trained, energized and strong advocates for all things steelhead. The pendulum reversed itself over the next 20 years, but at least we had a yardstick to evaluate options and their consequences. From a fishing perspective – and that's what these pages are mostly about – it was the perfect time to be there and be able to understand and influence that ever shifting baseline.

Whereas the best of what southwestern British Columbia once offered predated me, the promise of discovery was still alive through much of the magnificent northwest quarter of British Columbia in the mid-1980s. "Progress" was right on my heels, though. If it wasn't logging and mining or, more recently, pipelines, it was the gold rush of guiding. One had to see precious little rivers like the Damdochax and Kwinageese, the Tahltan and the Nakina, the Kluatantan and the Sustut/Johanson and know something of their fish and fishing to appreciate what development and people bent on exploiting oil and gas, wood fibre and precious metals achieve. And, about that guiding: Yes, steelhead in places such as these are valuable and marketable and the growing industry they support is certainly preferred over the extractive ones, but there are limits to the sustainability of the fishing being sold. Seldom are the players as altruistic as they make themselves to be. That too is part of my story.

The evolution of the regulations governing sport fishing for steelhead in British Columbia is another piece of mosaic that ought not to go unnoticed. Too few of the anglers and managers of the present are aware of that path. To that end I've added an appendix. For me, it illustrates the constant game of catch-up to try to match steelhead abundance to the sport safely available. If it doesn't also serve as a harbinger of seasons ahead we are not paying attention.

2

EARLY DAYS

Steelhead fishing, Lower Mainland style where I grew up, was highly competitive at the time. Anglers of the present might struggle to believe it, but there
were far more of their ilk lining riverbanks in the 1960s than there are today.
The difference is that then there were many more waters supporting fishable
populations of steelhead. The Vedder River was always the premier stream
that absorbed disproportionate effort, but the Chehalis, Alouette, Coquitlam,
Stave, Coquihalla, Silverhope, Capilano, Seymour, Cheakamus and a long list
of smaller waters kept that larger population of steelhead seekers relatively
evenly distributed.

Like most other Vancouver resident steelheaders, I spent most of my early
days afield on the Vedder. Cigarettes glowing in the dark on its prime runs
signalled which spots were reserved well before there was enough light to see a
fluorescent-tipped float dip from the take of a fish. The Easter Parade leaders of
the day used hollow fibreglass rods that had recently replaced their split cane
predecessors. Hardy reels were preferred. Neoprene was reserved for wetsuits,
not waders. Gore-Tex didn't exist. As unforgettable as all this Vedder experience was, that wasn't the Lower Mainland river that left the fondest memories. Those were reserved for the Squamish. Some of that relates to being old
enough to hold a driver's licence by the time I was introduced to the Squamish.
The independence that afforded was key to those 1960s adventures. My father
was thankful. Avoiding taxi service and sleeping in on weekends was fair
trade for borrowing his car.

The Squamish was relatively virgin territory in the late 1960s. That isn't to
say it wasn't a well-known steelhead river, but it was still a place where I could
spend an entire spring day with only myself as competition. There weren't
many days like that, but the midweek traffic was remarkably light, particularly
in the early part of a given season. The main tributary of the Squamish, the
Cheakamus, had a much longer steelhead fishing history. Its nerve centre was
Fergie's Lodge at Cheekye, roughly 2½ miles above the Cheakamus/Squamish
confluence. Fergie's began operation around 1910 and was into its third generation owner by the time of my first visit. The Cheakamus was readily accessible,

first by train and later road, and therefore more heavily fished. I loved the big water, though, and preferred searching the mother river for its fish before they were subjected to the gauntlet waiting upstream.

The Squamish was typical of all rivers arising in the mainland coast mountains. It usually stayed in the grip of winter until April's sun had the strength to cook enough snow to produce the midday pulses in flow that brought most of the annual supply of steelhead upriver. Depending on the year and snowpack, the fishing might be well underway in March, but the best of it was always in April and into May, runoff permitting. The Cheakamus could show fish early in the New Year and, sporadically, through decent weather spells in February and March. If they were encountered in the Cheakamus that early they obviously could have been had in the Squamish. True enough, but the onset of spring was always the best time to abandon North Vancouver and the Fraser Valley in favour of Squamish. Big snow winters would seem to be a thing of the past, but in my early years they meant delayed and limited access to some of the nicest water on the Squamish. The well-shaded rudimentary roads that wound their way in from the main logging road could hold snow for longer than one might think. Spring days were long. Even the youngest and most energized among us seldom fished from dark to dark, especially when on the river for more than a day at a time.

Tools of the trade circa 1970.

One feature of the Squamish's fish and fishing that continues to mystify me is the influence of the tunnelled diversion of Cheakamus water into the upper Squamish courtesy of the BC Hydro project that was completed in 1958. In effect that created two trails for Cheakamus-origin fish to follow on return to their river. In the pre-Hydro era there was no ready access and no sport fishery of record on those 15 miles of river between the Cheakamus proper and its new, secondary source at the powerhouse canal outlet. One is left to ponder how much, if any, of the fishery that developed on the upper

Squamish from the 1960s onward was just one more untapped opportunity always there or an artifact of the Hydro development. Two widely separated points for entry of Cheakamus water to the Squamish must have had some influence on fish behaviour as well as on patterns of steelhead production.

My years on the Squamish began about the same time the fishing opportunity on the upper river above the Cheakamus was entering its discovery phase. That time was sandwiched between my Vedder apprenticeship and departing South Burnaby in favour of another life on Vancouver Island. Wilson's Riffle, immediately upstream from whichever course the Cheakamus followed to reach the Squamish in any given year, was the source of my fondest memories.

Wilson's was a magnificent piece of water unlike any other on the Squamish. It came at the downstream end of a very long stretch of low gradient frog water that no self-respecting steelhead would bother to occupy for longer than it took to swim on through. At the time I came to know it I had no idea how its name was derived. Only recently did I learn that from Bob McKenzie, an old high school classmate and veteran steelheader. Friend Bob commands a remarkable breadth of knowledge on our cohort's association with BC's steelhead sport fishery. In the mid-1960s he worked at Harkley and Haywood, Vancouver's most revered fishing and hunting equipment store at the time. Stewart Wilson was the long-time manager of the store and Bob's boss. According to Bob, Wilson was a dedicated fly fisherman whose favourite piece of water on the Squamish was the riffle I came to know and love. Others familiar with Wilson and his frequent presence on that piece dubbed it Wilson's Riffle. To my knowledge Wilson and I never shared his run even though Bob's recollection was that there would have been at least some overlap in our respective years spent there. I'd be surprised if 40 years of spring freshets and some major fall floods haven't left Wilson's unrecognizable from what I knew, but I confess to being less than well informed on that point. It's been almost that long since my last cast there, although I'm told its name still stands.

The beauty of Wilson's was that there were endless places a steelhead might be found. There was at least 200 yards of optimal-velocity water with as much character and diversity as one could ever hope to find between opposite river banks. The near side was shallow but irregular and full of pockets well suited for a resting fish. The middle third was deeper but still thoroughly subscribed with ditches and pockets, each full of promise. Beyond, to the far bank, the river ran deeper and smoother over cobble and boulder substrate ideal for holding fish. At a decent April flow an entire day could be spent on the riffle without ever exhausting its potential or feeling it had been covered thoroughly. Local mythology had it that the fish came in on the tides and could show up at any time of the day. A fishless morning was never a deterrent. Wilson's narrowed at its downstream limit, picked up velocity and tumbled down to form a long, relatively nondescript piece that ran for several hundred yards to another fishy but far less attractive riffle. Halfway between, the Cheakamus carved constantly shifting paths through highly mobile alluvial gravels to meet the Squamish.

Reaching the multitude of pockets scattered throughout Wilson's was a challenge in itself. It wasn't a match for the Thompson's or the Campbell's treachery, but slippery, irregular cobble washed by waist-deep water commanded attention. Needless to say, you always wanted to fish the piece just out of reach of your best cast. Landing a fish, once hooked, meant a long and tricky wade back and down to the beach below. The experience could easily be wet. It didn't take long to figure out that some means of accessing the other side of the run would put me in a better position to do damage. My duck punt was put to use. That lasted long enough to convince me there were better options for crossing rivers. Next came a 9½ ft. Zodiac Cadet with a 10 HP prop outboard. Still later was my first jet pump, affixed to a 40 HP Mercury outboard mounted on the self-customized transom of a 14 ft. Zodiac Grand Raid. Now we were talking! Boats were the ticket, especially when yours was the only one on the river.[4] Staying ahead of the competition and getting to places no others could certainly paid dividends if fish was the metric, but the memories of those early days on foot were never replaced by the roar of a jet, no matter what the day's tally.

There were memorable moments before there were boats. One was the first landing of what most of us called a "big fish." Twenty pounds marked that threshold. (Somehow 9.1 kilograms doesn't cut it.) I'd never managed anything more than about 15 pounds until one of those early April mornings at Wilson's Riffle. I shared the water with one other angler for the first couple of hours that day and managed one averaged-sized fish before my company departed. Not long after, a good fish picked up my ghost shrimp and started backing downriver through the heavy water of the tailout heading for the long run bisected by the confluence of the Cheakamus River. My fish went slowly at first but picked up the pace the farther downstream it went. The standoff came at the confluence. That year the entire Cheakamus joined the Squamish at right angles through a steep, narrow chute whose loose and highly mobile substrate made crossing too risky. Immediately upstream the channel widened and the gradient flattened enough to invite the wade. With the fish already out in the middle of the Squamish methodically working its way downriver, there was no bringing it back to my side of the Cheakamus confluence.

Losing what I was convinced was the first "trophy" of my budding career because I couldn't or wouldn't follow it didn't make sense to me. Who would believe the story? There were no witnesses. I scurried upstream along the right bank of the Cheakamus, peeling off line until the backing on my Silex reel was near its limit, cinched my wading belt tight and strode boldly into the river. The alluvial gravel substrate didn't make for good footing, but I managed to stay upright in chest-deep water and stumble and lurch close enough to the far side of the Cheakamus to throw my rod up on the bank and clamber out. The fish was still attached when I retrieved the rod. The rest of it wasn't quite as dramatic but the fish did manage to pull me all the way to that next riffle, at least 800 yards downstream from where it was hooked, before I could beach him. By then waders filled with icy water mixed with nervous sweat didn't feel that bad.

I killed that fish. We were allowed 40 a year and no one thought there could ever be any conservation issue associated with filling a punch card. The professionals told us so. I promised myself it would be the only big fish I would ever kill deliberately. One didn't seem an unreasonable cost for ego gratification or satisfying the lust for an elusive 20-pounder. Besides, I had to prove that it had happened. I finally managed to pack that fish back to where I started and retrieve my vehicle for the trip to historic Fergie's Lodge on the Cheakamus, steelhead central in its day. Imagine my disappointment when their famous spring scale stopped at 19.5 lb. Suffice it did. To this day it remains the only fish of its size caught while angling outside taxpayer-funded operations that I didn't turn loose.

Years later there was another swimming event on the Squamish that was a bit more challenging. That came on a late April day, the last of a three-day trip. Again I was alone and the river was deserted. This time it was relatively high. There had been several days of warm spring weather with nighttime temperatures far enough from freezing that the overnight recovery of water volume and clarity was done for the year. I had launched my Zodiac and jet at the Mamquam River confluence well downstream from the best fishing water around the Cheakamus entry. It was hot when I returned to my vehicle at midday to load up and head home. There was a strong, warm wind blowing when I ran the boat up on the smooth sandy beach and made for my truck and trailer. I delayed long enough to grab a quick bite before returning to the boat to run it around to where I could get it on its trailer. As I reached water's edge the

Zodiac was drifting away a few short feet out of reach. The rising water and warm wind were just enough to push it off the beach and point it across and down the river. I watched helplessly as it found a new perch against a log jam on the outside of the bend a quarter mile downstream.

The 19.5 lb. fish from Wilson's Riffle, April 3, 1972.

Here we were again. Not a soul in sight, not another boat in the country, miles from anywhere and some big cold water between me and my cherished Zodiac. The solution was simple. Swim the river and retrieve it. I'd been around flowing water and done enough river swimming to understand what I was up against. It didn't seem all that risky. I got positioned well upstream, stripped down to my shorts, took a couple of deep breaths and launched myself into the flow. The instant cold pretty much anaesthetized any thought processes. All I can remember is that I needed to swim as far and as fast as possible to ride the thalweg to the boat. It obviously worked or I wouldn't be here to tell about it. It was much closer than I anticipated, though. I barely caught the rope strung along the gunwale of the Zodiac with a desperate lunge before being swept past and up against the logs below. I remember climbing up on the pontoon and lying there like a reptile until the sun's warmth restored some mobility and recharged a brain cell or two. One pull of the starting cord and I was back to where I had started an hour before. Things went a little more smoothly for the remainder of the trip home. My diary reminds me the fishing didn't match the rest of the adventure.

There were many other moments on the Squamish. One that stands out was the year when sampling of estuary substrate revealed high levels of mercury and aroused fears of contaminated fish that kept the punch card fillers at home. I believe it was 1970. The benefits were reaped by the few of us who didn't believe a fish as transient as a steelhead would ever bioaccumulate mercury originating from deep in estuary sediments. Besides, it didn't matter. We weren't the least bit interested in eating steelhead. Predictably, the days when the river was largely devoid of anglers and my favourite riffle wasn't reconfigured by the aftermath of logging didn't last. Sound familiar? Those gorgeous, black and white, clear-finned, sea lice–bearing creatures that came in bunches pushed by tides and hadn't lost their ocean feeding aggression were something to behold. There and in other rivers I came to know later there were no finer winter steelhead than those April fish.

Squamish River, May 2, 1975.

The 22 lb. Squamish River fish caught May 2, 1975.

3

PARADISE FOUND

In 1971 I departed the Lower Mainland in favour of Victoria to begin a career in fisheries management with the provincial government. A sharp eye on some of the dates in the previous chapter reveals that I didn't abandon my early history abruptly, though. Those occasional trips to fish the Squamish in the first few years of life on the Island might best be labelled transitioning.

My apprenticeship behind me, all I really needed when I took up residence in Victoria was a bit of advice from someone local to shorten the learning curve on this new (to me), river-laced real estate. It didn't take long to make a connection in the headquarters office of my new employer. Ted Harding Jr. was the man.

Ted came with a strong steelheading heritage. His boyhood home in Nanaimo was a frequent weekend haunt of some well-known pioneers of steelhead fishing in British Columbia, among them Dave Maw, one of the founding fathers of the Steelhead Society of British Columbia and its first president. Before that, Maw was a fish and game club champion and among the first executives of the BC Federation of Fish and Game Clubs (the forerunner of the BC Wildlife Federation). Someone once described him as the unchallenged dean of steelhead fishing. His adventures and highly envied success as a steelhead angler were the subject of numerous entries in the Vancouver daily newspapers.

Maw was a Dominion Bridge colleague of Ted's father, Ted Sr., albeit at opposite ends of a ferry ride. Ted Sr. and his circle of Vancouver Island resident fishing companions were equally dedicated charter members of the Steelhead Society. Ted Jr. referred to Maw as Uncle Dave. I heard many compelling stories about Ted's adopted uncle, his father and a long list of their angling companions who were a large part of the history of steelhead fishing on rivers such as the Stamp, the Ash, the Brem, the Thompson and the Dean, to name a few. Uncle Dave was a film buff too and had some incredible footage of steelhead fishing that he understood well enough to know a record was worth the effort. The generosity of the Hardings in providing weekend room and board for the crowd that was often present was over the top. Their Saturday dinner table could be surrounded by masters each of whom had logged years of steelheading at least half again my age at the time. To me the dinner table was a classroom. It was worth listening.

Between Ted Harding Sr. and Uncle Dave, Ted Jr. was moulded into a fisherman you didn't follow and expect to find leftovers. His preferred rod was a Hardy cane matched with a 3¾ in. Hardy Jewel. Ted was old school. No other fishing companion I ever had remained married to bamboo for so long after the appearance of new age fibreglass and graphite. Ted's habit of turning that cane rod over to equalize time spent with the rod bent on opposite sides while playing a steelhead was evidence of the teachings of Uncle Dave. He reasoned that a cane rod might develop a set from being bent only one way too many times. I couldn't have had a better companion to squire me around the best water on the southern Island. We fished many more rivers than just the Nanaimo while based out of the Harding residence, but it was closest and its most productive steelhead water was readily accessible. To say the fishery of today is a far cry from what it was in my introductory days hardly tells the story. The disturbing thing is just how quickly that decline unfolded.

In the 1970s a typical mid-winter weekend on the river reaches downstream from the Island Highway crossing would find several Vancouver-based anglers with campers parked at the forestry run on the north side of the river within sight of the bridge. The Friday evening ferries from the mainland were guaranteed to be well populated with tyros headed not just there but for the Cowichan, Englishman, the Qualicums, the Puntledge and Oyster and so on, all the way to Campbell River and beyond to the frontiers of Gold River and Woss. Sunday evening sailings back to Horseshoe Bay were equally subscribed with happy anglers anxious to share stories of weekend glory. The Vancouver dailies carried their weekly reports on all those rivers. The *Sun* accounts by Lee Straight and the *Province*'s by Mike Crammond were the equivalent of the Internet of today in terms of influencing the behaviour (and opinions) of the fishing public. No river within striking distance of a ferry terminal escaped.

Ted and I were of similar mind when it came to traffic and competition. We always sought the quiet places and times. He knew where to find them and how to keep them that way. Those Saturday dinner sessions at his parent's home astounded me in that respect. When everyone was assembled and the events of the day were expected to be shared, there was never a straight story from Ted Jr. His father would disclose full details on every fish encountered. Ted Jr., on the other hand, wouldn't give up any useful information for love or money. I was always caught in the middle. I figured Sunday fishing with Ted the younger would be compromised if I gave the straight goods on how many and where and Ted the elder was likely to take umbrage if I didn't. Eventually I figured out that the chess game had been going on for years and had come to be an accepted part of weekend dinners.

Staying away from traffic on a river as popular and accessible as the Nanaimo was not always easy. Ted and I had an ace in the hole, though. That water wasn't exactly obscure or inaccessible. It was barely out of sight looking upstream from the highway bridge, but the canyon walls around the bridge (now the Bungee Zone) meant you couldn't just walk up the river to get to water more attractive or fishable. Then, even if you did know how to gain access through the Hub City gravel pit to get back down to water's edge, there wasn't much to fish before the river was completely bounded by steep canyon walls for miles upstream. That seemed good enough to keep all the highliners from Vancouver concentrated on the more attractive river reaches downstream from the highway.

The gravel pit was kind to us almost every time we went there. Of course the water had to be at the proper height or we didn't bother. It was a long, straight piece constricted by bedrock banks. High discharge meant too much velocity and nowhere a fish was likely to hold. At low flows the reach was too slow to fish

The "Gravel Pit" on the Nanaimo River, 1972.

effectively. There were several perches from which to make a cast, and most days would see a fish for one or the other of us from each of them. There was one spot, however, that consistently produced multiple fish. It wasn't one my friend Ted had spent time at before, and it came as a surprise when I did.

Tight against the far bank at the tail end of the stretch we liked to fish there was a short little break in the otherwise smooth surface that led to it from several hundred yards above. That spot was completely inaccessible from that side and required a fairly long cast and carefully managed drift with a float rod to run a bait through it properly from the near side. Neither of us had ever bothered with it previously. On one of those days when we had finished with what we always considered the prime water upstream, I decided to try something different. As I recall, it was one of those "oh, what the hell" decisions. There was nothing to suggest anything special about that little break in the surface. If anything it looked like it might be a shale outcrop only a few inches below the surface that had created it. I think I was more worried about hanging up and breaking off a leader than finding a fish, but it would only take one cast to tell. I waded out and dropped one as gently as I could at the edge of the break just above its start. The float disappeared almost immediately, and the first of many fish that came from precisely that same two or three square metres over the next two years was soon at hand. Ted and I often consumed most of the time spent driving back to Victoria on Sunday nights theorizing about the factors that made that one little spot so productive.

Everyone who ever stayed with fishing a good river long enough to find its fish consistently probably had a favourite spot. I already spoke of Wilson's Riffle on the Squamish as one of those. This tiny pocket on the Nanaimo was different. There was never any searching or mystery about it. If the river was anywhere near the height preferred when venturing into our gravel pit stretch, and if the cast was made where it should have been, it never failed to produce at least one fish. My best session there was five good fish to the beach and a couple lost. There were many twos and threes. It was a rare event to not hook a fish on the day's first cast to that magic spot. Forty years later I've found only one other spot I'll speak to later that came close to providing such consistent results over any extended period. When I read Ralph Wahl's wonderful little 1989 book, *One Man's Steelhead Shangri-La*, memories of my Nanaimo pocket were born again.

Steelhead catch success as I had known it out there in the Fraser Valley in the years before Nanaimo could be signalled with the fingers of one hand most years. Now a day's catch often exceeded a season's total, and a day without a fish bordered on unimaginable. Such were the times. As an expatriate from the highly competitive Lower Mainland fishing closer to my birthplace, I took full advantage of everything I now understood I'd been missing. The drive to make up for lost time was all consuming, and the Nanaimo bore the brunt of that in the early going. Zeke Withler, my boss in Victoria and a highly respected mentor, was well aware of my obsession. I'll never forget his words on one of those Monday mornings following another intense weekend of fishing. I looked up from my desk to see his head sideways in my office doorway as he said, "Hooton, if Saint Peter's a steelhead you're in big trouble."

Good fishing on the Nanaimo wasn't restricted to the gravel pit area in the early 1970s. There were many good days on reaches downstream from the highway crossing. March was the month to really fatten the averages and, as with so many other rivers that tended to fill with weekend warriors, you wanted to be there on some of those other five days.

In retrospect the early 1970s was the time against which I measure the Nanaimo and many other Island rivers. Even by 1975 when I had moved to Nanaimo, the fishing experienced during my introductory period short years before was slipping noticeably. The pocket at the tailout of the gravel pit run looked the same as it always had, but any predictability of results had vanished. Subtle changes, undetectable from a surface view, may have altered it such that it was no longer the stopover point for every passing fish it seemed to be when first discovered, but fish supply was the more obvious explanation. There was still the occasional productive, pre-work morning in March of 1975 and 1976 but the best of it was already behind me. Greener pastures beckoned.

It's hard to completely abandon a river as kind as the Nanaimo was to me all those years ago, but there was a long pause. Thirty years passed between casts. I've kept in touch with trusted confidants, though, and I know they still catch the occasional fish from reaches that we managed to convince bureaucrats could be reopened to angling after a long period of conservation-predicated closure did nothing more than eliminate what little legal angling opportunity was otherwise available. I still wander along some of those reaches myself, my pup in tow and perhaps a new fly line to test for serious application elsewhere. As if to torment me, an early 2014 fish rekindled memories of misty spring mornings with the first rays of sun streaming through white-barked streamside alders, air heavy with the smell of skunk cabbage and trilliums all along the shaded paths. I'll not forget that fish for inspiring those reflections and reminding me how resilient these wonderful animals are given half a chance.

An April 2014 fish from the Nanaimo River.

4

A DAGGER TO THE HEART

My first encounter with the Englishman River came on a family camping trip that found us at Englishman Falls Park in the early summer. I'm guessing it was 1956 or 1957. The only thing I really remember about that trip was the fish I saw each morning. There were three or four of them. They were huge, grey, ghost-like creatures hovering at mid-depth along the edges of the canyon walls underneath the bridge that crossed the lower falls pool just before it spread out into a broad, cobble-lined tailout and began its tumble toward Parksville Bay ten miles distant. Long afterward I came to appreciate that those fish would have been kelts but at the time nothing like that mattered. The seed had been planted.

Fifteen years later my very first cast into the Englishman River was rewarded with a shiny new steelhead. It was early on a Saturday morning after one of those Friday night drives from Victoria to Chez Harding in Nanaimo. Once again my guide and companion was Ted. He had been telling me for weeks about his experiences on the Englishman and promised to drag me along one day when the time and water were right. This was it.

The Englishman is small enough to be dependent on freshets to produce its best fishing. At a good fishing flow the river couldn't be crossed, at least not by wading. The ideal circumstances usually followed a rainfall event that spilled enough sky water to swell the flow but not so much as to put the river in the trees for several days. Either that or you waited out the freshet to make your move at the preferred stage on the descending limb of the hydrograph. Needless to say the winter storms that brought the river into prime fishing condition didn't pay heed to the day of the week, so it wasn't just any weekend that one could descend on the water and find optimal conditions. On that first visit my man Ted knew we were on the money. It was a good day to be introduced to what quickly became one of my all-time favourite rivers.

The Englishman was as near to an ideal steelhead stream as one could find on the east coast of Vancouver Island in the early 1970s. Granted it had long since exhibited much evidence of unravelling due to the cumulative effects of logging in its headwaters but, at that time at least, it was still a joy to fish. It was a higher, more uniform gradient stream than many others north and south. The runs and riffles were cobble lined. Shale outcrops that made for unappealing and unproductive stretches of water on some neighbouring streams were rare along the Englishman. The river was well defined and its holding water easily read. It was never a salmon stream of any significance in the years I knew it. Yes, it still had a few hundred chums in the best years, occasional very late run coho (my partner Sean Hay once caught a newly arrived female near Englishman Falls in the first week of February) and, rarely, a stray chinook, but no more. It was trout water through and through. Excluding the few years of significant chum presence, steelhead outnumbered salmon. That may not have been the case decades before but it certainly was by the 1970s. How many winter steelhead streams exhibit that feature?

There was only one road in to the reaches of the river of greatest interest to legitimate anglers, so it was never hard to gauge traffic levels and where the competition might be. In those first few years that road was in pretty rough shape and kept traffic to a minimum. The usual fare was to park and walk the river, either from the top to the bottom of the best fishing water or from the downstream end up. The Claybanks near the South Fork confluence downstream to what we called the Grassy Banks just above the old Top Bridge area (now a regional park) was the preferred beat. It was only 2¼ miles from end to end, but on the good days the number of fish encounters could keep a proficient rod from fishing it all thoroughly.

On one of those early days when Ted and I were hiking and fishing our way down the river we decided to go farther than we normally did. That took us a bit downstream of old Allsbrook Road that terminated at Top Bridge. The river passed though a short narrow canyon at that point and a bridge had once been built there. Local lore had it that the bridge had succumbed to high water following one of those rain-on-snow winter storms several decades ago. It wasn't far below that point where the river gradient declined and its definition and appeal for angling diminished. I recall the last spot we fished that day was created by a classic old growth cedar that had fallen into the river years before. Its twin spires were anchored by its massive flared root complex still firmly embedded in the stable bank at river left. The root structure created a break in the river's velocity and some quieter, deeper water all along the edge of the two spires. It wasn't enough to offer anything other than temporary sanctuary to passing steelhead but it was a bit of a feature of an otherwise nondescript stretch of river. At the downstream end of the spires the riparian owner had built a small platform whose primary purpose looked to have more to do with dabbling feet on a hot summer's day than fishing. The platform was old and weathered. That spot stuck in my mind as a point of reference. In 1978, six years later, my new wife and I bought the property defined by those cedar spires. From that point forward my story of the Englishman is not as much about its fish as it is about its habitat. What happened to that wonderful little river in the space of the nine years we lived on the property whose boundary was the centre of its channel was heartbreaking.

The Englishman is just one of dozens of formerly well-known and revered steelhead angling streams on Vancouver Island victimized by logging. By the time the first of the new age freshwater fishery managers and the environmental community arrived around 1970 all the best valley bottoms had already been clear-cut. There were no rules around road locations or standards, stream crossings, drainage structures, rates of cut, leave strips etc. Short of the most obvious and deliberate actions such as removing gravel from stream channels for road building purposes or falling and yarding timber into and through flowing water, it was no holds barred. Tales of explosives thrown into pools full of fish and logging camp occupants with no respect for regulations or limits were never hard to uncover. The cumulative effects of decades of those practices were bad enough but the serious damage had yet to unfold. That came when the biggest, best and most accessible timber had been liquidated and the thirst for fibre took the loggers to the steep slopes and the highest elevations to harvest what timber remained in the back ends of all the side valleys. The cumulative effects are ongoing and will continue to test the resiliency of steelhead populations as never before.

In those late 1970s and 1980s years we lived on the Englishman I spent as many days climbing its steep headwater slopes in search of elusive blacktail bucks as I did walking and wading its lower reaches harassing steelhead. It didn't take long to appreciate how little old growth timber remained in that watershed and how quickly it was disappearing. Observing sub-basin after sub-basin denuded of timber and left with no capacity to absorb and filter coastal rain storms taught lessons never forgotten. Hydrologists of the moment constantly debated the issues. The forest company men were constantly toe to toe with government fisheries specialists. New constraints around road building and drainage structures on steep slopes, on stream crossings, on rates of cut and clear-cut size and, occasionally, even on streamside buffer zones emerged from some of that. Fifty or 75 years earlier might have made a difference, but by the last decades of the 20th century the unravelling of the river channel was irreversible.

If I had to point to a single event that sent the Englishman steelhead habitat and, ultimately, its fishery on the downward spiral it has never begun to recover from, it was a classic warm, wind-driven rainstorm that occurred shortly after we took up residence on its bank. It was the winter of 1978–79, an unusual one by any metric. There were good numbers of steelhead in the river by mid-December that season.[5] Those in the know were well aware they could stand on the highway bridge entering Parksville and see a dozen or two steelhead holding in the shadow of the rock ledges on river left immediately upstream. A similar pod of fish occupied what we referred to as Hole in the Bend, a few hundred yards downstream from the South Fork confluence. The other thing that stood out that winter was the cold. The river began to freeze over by Christmas, and by New Year's all the holding water deep enough to hide fish for any prolonged period was sealed by ice thick enough to walk on. Anchor ice just kept building in all the riffle areas, creating new courses daily for what little free-flowing water still remained at the surface. The weather event that catalyzed the steady demise of the river channel came in late February 1979, not long after the thaw.

Big rain, warm wind, quickly receding freezing elevation and a snowpack no longer protected by a mature forest canopy is not a happy story for any river. Valleys and rivers that had sustained winter storms for centuries but remained relatively unaltered reached a breaking point after humans entered the equation. When the heavy rain began on February 24, the buffering effect of old growth trees on slopes with natural drainage paths to spread the release of water to the valley below was no longer there. Too great a proportion of the Englishman Valley had been stripped and too many steep slopes criss-crossed repeatedly by access roads. Each of those roads intercepted water from barren slopes above. With no mature trees to act as the sponge and natural water courses no longer capable of withstanding not just more meltwater but vastly increased concentration of it in the ditches that suited logging road builders, the predictable occurred. Ditches became raging torrents. They sent incalculable volumes of eroded soil and rock and logging slash pulsing down the slopes into the larger channels that ultimately converged to become the Englishman proper. The combination of sludge and trees grinding its way downstream during the February 24–25 storm destabilized and reconfigured the Englishman's steelhead water as never before. That was the beginning of the end for the capacity of the river and its fish to absorb the habitat abuse.

Ric Olmsted (RIP), Englishman
River, December 28, 1978.

One major flood event is serious business for steelhead and steelhead habitat. Unfortunately that February 1979 event proved to be only the first of a succession. There was another on Boxing Day in 1980 and a third in mid-November in 1983. The first flood completely reconfigured the prime steelhead fishing reaches between the South Fork confluence and Grassy Banks. Every piece of water that had remained essentially unaltered in the years between my first Englishman steelhead and the onset of winter 1979 was gone. One might assume you get one back for every one lost, but that was not the case.

By 1984 the net reduction in suitable holding and fishing water was dramatic. All the prime runs had been cemented. Silt, sand and fine gravel had replaced cobble and boulders in run after run. Where once there were runs that would hold a fish over most or all of their length, there remained only a few isolated patches of suitable fishing water in even the largest of them. Debris littered the banks, most of it freshly deposited by the most recent flow event and destined to be replaced by the next one. The river channel had become so unstable one had to relearn it after every significant freshet.

The ultimate barometer of the state of the Englishman was the number of steelhead it supported before and after the destabilization that began in early 1979. The fish that returned that year and over the next two winters would have been at sea and unaffected by the state of the freshwater habitat they would encounter on return. Their offspring didn't do so well. I like to recall the snorkel surveys I supervised in those years and one particular beach seining operation undertaken in the late winter of 1978–79. On that occasion my Fish and Wildlife Branch crew captured 126 steelhead in one set at the Claybanks Pool immediately downstream from the South Fork Englishman confluence. At least three of those fish exceeded 20 lbs. My successors tell me their maximum steelhead counts over the past dozen years have ranged from about 20 to 100 and averaged around 50. That's for ten miles of river, not a single pool. The days when I hiked down to the Englishman Falls plunge pool at daybreak in early spring and observed 100 or more steelhead quietly resting in the broad, glass-smooth, cobble-bottomed tailout are another piece of history. The present cast of players who are convinced they know something about the Englishman would seem to regard such information as fiction.

Proving the obvious to the satisfaction of those in a position to alter the status quo is always an exercise in frustration. If ever there is a constant in the world of fish and fish habitat management that is it. No amount of data from streamflow records or well-documented changes in stream channel location and configuration has ever been sufficient to motivate change. Steelhead simply don't count. The only thing left to do is to try to leave a record for those who might find such things instructive. Pictures do more justice to the story than words.

"Grassy Banks," Englishman River, January, 1979, 1981 and 1983.

Top: "Top Bridge," Englishman River, illustrating river level baseline.
Bottom two: "Top Bridge," Englishman River, illustrating flood stage river levels.

Upstream and downstream views of the old growth cedar spires and platform built between them at the author's property, Englishman River, late 1979.

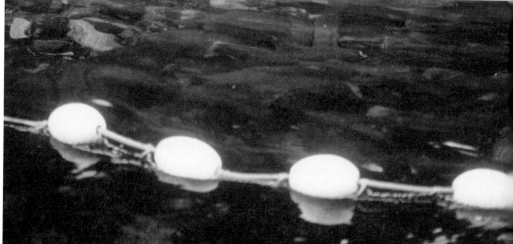

The beach seining operation and some of the 126 fish captured in one set at the
"Claybanks," Englishman River, April 1979.

The "Claybanks," Englishman River on February 5, 2012, photographed from exactly the same spot as the opposites page images.

There is a postscript due regarding the state of the Englishman River today. The entire river was closed to angling during the steelhead season in 1999 and, save for the relatively poor fishing reaches downstream from the old Top Bridge which were reopened to catch-and-release only in 2008, has remained so through to the present. This was one of several such closures invoked on east coast Vancouver Island streams at a time when the estimated adult steelhead populations in each of them had plummeted to a few dozen at most. The combination of reduced capacity of these logging-impacted watersheds to produce steelhead smolts and modern era lows in ocean survival rates for those that made it to that life stage was assumed to be the problem. It was speculated that investment in improving the freshwater habitat conditions to augment smolt production would offset the trough in marine survival rates and avoid looming extirpation. Fifteen years and hundreds of thousands of dollars later a few outcomes seem obvious, at least to me.

Fixing devastated stream habitat on a scale grand enough to demonstrably increase smolt production is a very tall order. When the Englishman River became destabilized to the point it did as the cumulative effects of too much clear cutting, too fast, of its upper slopes manifested themselves through the 1980s and beyond, the fix became impossible. The reality is that the number of steelhead smolts produced would have had to increase several fold to offset marine survival declines. Conversely, a very minor increase in marine survival would increase the number of adults returning from smolts of a given brood year regardless of whether or not any manipulation of the freshwater habitat had occurred. No one in the management community has produced any evidence that the number of steelhead smolts increased as a result of work done, nor has there ever been any plausible linkage shown between any of that work and adults observed during snorkel observations over the past decade. Feel-good it is, fish producing it isn't.

The other noteworthy element regarding all the investment in steelhead habitat improvement on the Englishman is who paid for it. The logging companies involved have changed over time as their original holdings were sold and resold. Those largely responsible for the legacy of ruined fish habitat now ply their trade in the rainforests of South America. Fisheries agencies have never had power or authority to change how business was done in the Englishman drainage, but it is particularly offensive that they found themselves having to pay the logging companies for the equipment, materials and riparian access development required to pursue the stream improvement initiatives. Imagine purchasing root wads and cable from the industry responsible for uncountable numbers of the former ploughing their way down the river during freshets and numerous lengths of the latter left in the bush in bygone years.

The Englishman has not escaped the ubiquitous band-aid of hatchery intervention. The off-channel habitat improvements, to be fair, have greatly increased the coho smolt–rearing capacity of the lower river system. Whereas the salmon-to-steelhead ratio favoured the latter in the 1970s and 1980s, that is no longer the case. A hatchery now takes responsibility for ensuring all that nice new off-channel coho habitat is utilized. Natural spawning in the channel itself is probably minimal due to the dearth of suitable gravel substrate. That keeps returning adults out in the main river where they are at the mercy of all the forces that inspired the creation of the off-channel rearing. Thus the hatchery. As long as the hatchery program continues and the ocean doesn't get any more inhospitable the coho returns should hold up. I often wonder if anyone ever considered that water diverted from the mainstream river to supply the off-channel coho-rearing habitat doesn't help steelhead trying to hang on in the river itself.

Lastly, those angling regulations. Fifteen years of closure have demonstrated only that elimination of anglers is counterproductive. No more steelhead can be shown to have resulted, even with all the investment in habitat work. No anglers equals no voice to make a case against developments that continue to erode fish habitat and fishing opportunity on the Englishman and elsewhere. Who even notices? The managers have been painted into a corner that forces them to be more concerned about their own survival than anything else. Steelhead angling advocacy is a distant memory. What was once a rough logging road that provided the access for those of us who frequented the Claybanks to Grassy Banks reach on trails created and used exclusively by anglers now sees every manner of outdoor enthusiast except anglers. Hikers, dog walkers, horse riders, mountain bike enthusiasts, naturalists, summer sun seekers – all are well represented. The interpretive signs, the trail markers, the benches, the bridges and more that come with the regional park and protected area status for the areas once known only by the fishermen who put the river on the public radar originally are all good news. Sadly, however, the steady parade of users of the area today know nothing of how it all began.

5

PARADISE LOST

The list of British Columbia steelhead streams worthy of diamond button[6] status is short, and of those that were not already well known by 1960 even shorter. Gold River was one of them. That its entry into the public angling domain lagged behind every other major steelhead producer on Vancouver Island is due strictly to the access policies of logging companies who controlled the roads that crossed the Island between Campbell River and Gold River.

The best angling-related historic references to the Gold River come from Lee Richardson's limited edition 1978 book *Lee Richardson's B.C.: Tales of Fishing in British Columbia.*[7] Therein he recounted his experiences flying into Gold and Muchalat lakes in 1954. The forests of the lower Gold and Muchalat were described as the most magnificent stands of virgin timber remaining on Vancouver Island, rivalled only by those found in Washington's Olympic National Park. A quote from Richardson serves as illustration:

> We were fortunate to behold this great forest with trees so tall they were lost in clouds, trees that were old when Columbus discovered America, many of them eight feet at the butt and without a limb for a hundred feet. The earth below was covered with golden moss, like a vast Persian carpet. Sword ferns grew in bright green clusters beneath the trees, and little else. Here time had dealt kindly with the land, which until recent years had known only the red men, and few even of them. Standing in the shadow of these giants one knew them to be the handiwork of the Creator, a great cathedral that would never be duplicated.[8]

Richardson also made note of the progression of logging access up the lower Gold from tidewater. That was underway at the time of his 1954 visits to the lakes above. Eleven years later the blasting of a road through the lower Gold canyons by the East Asiatic Company was complete, there was a pulp mill in the river's estuary and the instant company town of Gold River, 2,500 strong, had been established eight miles upstream, right at the confluence of Haig-Brown's beloved Heber River. Thank you for the benchmark, Lee Richardson.

There are a few additional bits of information on early arrivals to the Gold River steelhead scene and how they got there. Luhr Jensen, the renowned fishing lure manufacturer from Hood River, Oregon, reportedly flew to the river's estuary in 1960 and was accommodated by logging company officials once there. Lee Straight spoke to the public access problems in a January 1961 column where he mentioned having visited the Gold River three times by then, always by aircraft. He complained bitterly about government intransigence in addressing privatized logging roads on public land. Straight also recounted a salmon fishing trip he made in August 1962. On that occasion he drove to Strathcona Lodge on upper Campbell Lake (still in operation). From there he reached Gold River courtesy of a short boat ride across the lake to meet the regularly scheduled Island Coach Line bus which was sanctioned to use the Elk River Timber Company road that paralleled the opposite shoreline to deliver people to the distant Gold River and Thasis logging camps. The Strathcona Lodge owner had a salmon fishing camp at the mouth of Gold River and special arrangements with the loggers to get his clients from the bus stop at their camp down to his facility. Another reference to the early years comes from Boston resident Edward Weeks,[9] who accompanied Roderick Haig-Brown, his neighbour Maxine Van Egan (RIP) and Skate Hames (conservation officer, renowned cougar hunter and close friend of Haig-Brown) on a two-day steelhead fishing excursion. The precise date was not specified, but one can interpolate from his descriptions that it was around 1960. Weeks noted that they passed through seven gates to get to their fishing location. Hames, as the designated predator hunter for the area, had keys to the gates. It wasn't until June 1963, when the government passed a new Access Act, that the general public was finally allowed passage beyond the company gates during non-operating hours. So began the exponential growth in fishing pressure.

My personal association with the Gold River began with a trip through the area at the Christmas break during my first year of university in 1965. A couple of us decided to head to the Nimpkish to try to cash in on rumours of steelhead aplenty. Getting to the Nimpkish was a bit different from what it is today. One had to cross the Island by logging road from Campbell River to Gold River (during non-industrial hours of course), parallel it upstream for several miles and then cut diagonally back across the Island to the logging community of Woss, centrally situated on the Nimpkish. The access to the two watersheds, Gold and Nimpkish, was governed by different logging companies, and there was a gate and a gatekeeper who monitored the traffic between them. The gate we had to pass through was at Oktwanch, a tributary to Muchalat Lake, the primary producer of the Gold's sockeye, as well as a fair proportion of its winter steelhead. The conversation we had with the gatekeeper is permanently etched in my brain. Among other things he told me his roommate at the logging camp in Gold River had caught nine steelhead in the Oktwanch River to date that winter. I'll come back to that.

A second trip through Gold River came in the spring of 1972. By then I had begun my career in the fisheries management world and was on my way to northern Vancouver Island on a fish habitat–related issue of the moment. I organized my travel to be able to spend an evening and the following morning trying to catch my first Gold River steelhead. My choice of water to fish was Egan's Run, immediately downstream from the Muchalat River confluence. Years later it became known as Branch 14 but, to me, it will always be Egan's – as in Roderick Haig-Brown's neighbours and fishing companions Van and Maxine.[10] It was as perfect a piece of steelhead water as I had ever seen – 150 yards of prime depth, speed, bottom configuration and holding water from one side to the other. It was still only mid-May, and early to be expecting the first of the season's summer steelhead, but I did manage one landed and another lost before resuming the business trip north. Egan's was the second indelible impression I'll come back to.

My Gold River story is not is unlike that told for the Englishman. It's more about loss than glory. First, though, it's worth the ink to speak to the Gold as it was and its status in a broader context. I don't hold opinions about the Gold River without justification. Well over a half century of experience of catching steelhead in 81 rivers in British Columbia, including the Dean, the Thompson, a plethora of streams on Vancouver Island and Haida Gwaii, every big name Skeena tributary and a host of other lesser-known rivers throughout the steelhead range in the province allows me that perspective. In its prime the Gold was pure steelhead habitat from the lake at its headwaters all the way to its estuary. Unlike almost all the other coastal streams that typically display high gradient upper reaches, moderate gradients in their middle reaches and flat, braided channels strewn with woody debris in their lower reaches, the Gold's gradient was relatively uniform throughout. Its substrate was cobble and boulder even in the broad tailout of the lowermost run flooded daily by ocean tides. Gravel substrate was scarce, present only in occasional patches along an edge or behind a larger boulder. Sand and silt were as scarce as gravel. The dearth of smaller substrates had us biologists speculating that the river was spawning habitat–limited. The clean cobble- and boulder-lined runs that were a dream come true for fish were as prime juvenile steelhead–rearing habitat as ever I saw on Vancouver Island. And then there were the fish.

There were other rivers in British Columbia that grew more steelhead than the Gold and there were many with larger fish. What separated it from all those others I became familiar with was the opportunity to find steelhead fresh from salt from November through July. The other three months could have produced the occasional new fish but the expected rewards at that time of year were never enough to invite me to address the question thoroughly. A July steelhead from the lower Dean or Skeena is something to behold. Of that there can be no dispute, but having sampled the bounty of both I'll argue that, pound for pound, a late May Gold summer run is (was) as feisty as they come. The winter steelhead, though encumbered by water temperature and gonad development, were the equal of winter fish anywhere.

My first serious summer steelhead trip to the Gold was in the last week of June in 1972. A multi-day break from working on pulp mill issues in the Kootenays came at just the right time for me to get back to the water I'd briefly sampled six weeks earlier. Rumour had it that the best place to be was the Heber confluence. There was a wonderful little campsite immediately across the river from the Heber junction. It was my base for the next three days. Anyone visiting there today would likely find it hard to believe the existing industrial park was once a grove of magnificent old growth cedars and firs. The fishing proved beyond expectations built mostly around the apprentice-ship years back on the mainland around Vancouver. In three days on water I had never seen before I landed exactly 30 prime summer steelhead while never encountering another angler. I had no reference point at the time, but I learned eventually that the water conditions on that trip were as near to perfect as ever they are. Later years taught me it wasn't common to have such good water that well along in the season.

Not all my time on the June 1972 trip was occupied by the Heber conflu-ence pool. Although it wasn't easy to walk away from some of the best steel-head fishing I had ever seen to that point, it made sense to do some exploring. River right below the confluence was impossible to walk, but the side opposite sported a baseball diamond and running track tied to the school just across the Heber via foot bridge. A road from the townsite crossed the Heber slightly upstream to provide vehicle access to the area. Off I went, soon to discover that the road to the ball park continued on downstream. The trick was that it went straight through the town's (unfenced) garbage dump and every black bear in the country was tuned in. I counted eight in close proximity as I wandered past armed only with a fibreglass wand. The surrounding trees were alive with bald eagles and ravens anxious to announce the presence of an intruder. Slightly beyond the dump was the town's sewage treatment plant. Between garbage and sewage the experience wasn't much for aesthetics but there was a run at the end of it all that made the hike worthwhile. One of my largest Gold River summer runs came from there a year later.

The other area I explored on that early trip was the water that could be seen looking down from the road about halfway between the town and the pulp mill. I learned later that the most attractive run viewed from that point was locally known as Timmons'. Tim Timmons was a part-time guide who worked at the local hotel, the Gold River Inn, in the years immediately following its opening as the instant town's only accommodation facility. A few years later Timmons' became known as Clark's, as in Bill Clark, a well-known steelhead chaser from Campbell River. Access to Timmons' was by trail from a small pullout on the road high above. The trail was a joy to walk. It wound its way down through a beautiful patch of old growth forest of the kind described by Lee Richardson. Halfway down it forked – left to continue on to Timmons', right to the Crazy Hole. I took left and found myself on what I would come to know as one of the best and most consistent winter steelhead producers on the river. I caught summer fish there that day and in all the likely looking spots upstream. Into the memory bank it all went.

The trail to Timmons' became another story in itself. I believe it was 1977 when local authorities decided it would be a good idea to divert a natural watercourse that created a small waterfall off the rock cliffs near the point on the pulp mill road where one could view Timmons'. Spray from the falling water allegedly created a driving hazard in cold weather due to ice formation. A culvert that passed under the road adjacent to the falls allowed the creek to follow its natural course carved through the steep, solid rock embankment to where it entered the Gold just upstream from Timmons'. The redirected water was run through a newly installed culvert that emptied onto the river side of the road about 150 yards downstream. From the culvert outfall the water found a path of least resistance to the river below. Those involved were oblivious to the consequences of pouring water onto a soft forest floor hundreds of years in the making. One freshet later the erosion was unbelievable. Trenches ten feet deep and ten feet wide were gouged all through the grove of old growth trees.

The roots of a dozen or more of them were quickly undermined, whereupon they fell like pickup sticks across the yawning ditches. So much for as nice a trail as ever graced the Gold.

There were other memorable moments with summer fish in the Gold. My friend Ted Harding and I shared a wonderful May 24th weekend exploring the lowermost reaches of the river in 1974. That was our first exposure to Muchalat-bound sockeye. They passed us in waves on the incoming tides and moved steadily upriver. It was the only year in at least a decade of visiting the same times and places that I witnessed sockeye in such abundance. They broke the surface constantly with their characteristic flips until well after the flood tide peaked. Among them was the occasional summer steelhead, clearly distinguishable by its larger size and its trademark porpoise-like rolls. It always frustrated us that the ratio of sockeye to steelhead was so heavily skewed to the former while only the latter could be harvested legally. On one of those evenings, however, Ted caught four sockeye in rapid succession while fishing roe in tidal water. A couple of those delectables went back to Nanaimo.

Experience taught me there were two distinct stocks of summer steelhead in the Gold. The upper Gold fish, much larger on average, came earliest. Those were the fish whose timing seemed linked to the Muchalat-bound sockeye. The Heber fish, always smaller, came later and weren't anywhere near as obvious in the lower river reaches. They seemed to move quickly to the Heber confluence area. These observations were not speculation. They were wholly supported by annual snorkel surveys I initiated with my fisheries staff in the years following my 1975 instalment as the province's regional steelhead biologist in Nanaimo. For whatever reasons the fish we counted in the upper Gold always included a complement of the older, larger fish that were almost never seen in the Heber. Fish clearly in excess of 12 lbs. were not abundant in the upper Gold, but there were always some.

In those bygone years of the early season fishing in the lower river there was one truly exceptional fish worth mentioning. It came on another of those May long weekend excursions, this time in 1976. I had been fishing with office mate Doug Morrison for the first two days of the trip, and we had managed a few of those magnificent early timed fish. Doug had to return to Nanaimo earlier than I did and left the river to me on that last day. I started at one of the spots that had produced fish the previous two days and was into a good fish immediately. It didn't do anything spectacular nor did it resist for all that long, but I knew it wasn't an average fish. Ten minutes on I found myself staring down at a nickel bright male, the largest coastal summer steelhead I had ever seen. I packed a precision Salter spring balance and custom-sewn weigh bag in my vest in those days. The balance went to 22 lb. It bottomed out when I placed the fish in the bag and lifted it. With no camera to put that fish on record all I could do was remove a few scales and send it on its way. I made a special effort to process a select scale and make an acetate impression suitable for a colleague at the Pacific Biological Station to photograph and prepare an 8 × 10 print. The framed product still hangs on my office wall as a reminder of a once-in-a-lifetime fish. Over all the years of fishing the Gold and all the snorkel surveys undertaken on both the Gold and the Heber rivers I never saw a summer steelhead that I was certain would come within four pounds of that fish.

Summer steelhead fishing showed no signs of change for those first few years. The river was unaltered from year to year and the traffic exceedingly light. I often wondered how good it had been in the years before I arrived. Fragments of information cobbled together instruct me that there were at least three guides operating out of the Gold River Inn and as many as five more from elsewhere who targeted the Gold's summer and winter steelhead. There were no reporting requirements for guides at the time and therefore no record of how many fish they removed. All that can be said is the catch and possession limits at the time were exceedingly liberal (see Appendix) and there was no such thing as catch-and-release. In spite of the relatively good fishing I experienced in my early years, the complete absence of guides by the time I arrived, only seven years after the opening of logging gates and the birth of the town, implies the fishing had slipped significantly.

Along with the disappearance of guides, most of the steelhead hawks from Campbell River turned their summer attention to the emerging game in town – mooching cut plug herring for chinook in the tidal back eddies of Discovery Passage and Seymour Narrows. I succumbed to that myself in the late 1970s and left the Gold summer runs alone. A watching brief was maintained, however. That was enough to reveal that the late May encounters with those magnificent summer fish were a fading memory by the late 1980s. The mystery of it was that the comparative snorkel counts in the upper Gold later in the summer never reflected the markedly reduced angling results in the lower river months before. I have no explanation for that. Finding a late May summer run steelhead in the lower Gold today is harder than it ever has been.

Winter steelhead were the bread and butter of the Gold's reputation. By late November the hard core outdoors enthusiasts of the 1960s and 1970s were done with their ocean fishing, hunting seasons were all but over and the skiing at Mount Washington wasn't the draw it became in the 1980s. Steelhead fishing was the only game in town. The Thompson kept some Lower Mainlanders occupied a bit longer, but all the popular Island rivers of the day – the Campbell, the Quinsam, the Nimpkish, both Qualicums, the Cowichan, the Nanaimo and so on – were well subscribed with anglers in December. The Gold was widely considered the best of them all. Creel surveys documented how busy it was between Christmas and New Year's, when many Vancouver, Campbell River and lower Vancouver Island residents frequented all the best runs. There weren't very many residents of the town of Gold River who posed any competition for the highliners from outside until the best years on the Gold had already become a memory.

Winter steelhead fishing was a numbers game. Until 1979, when the first of the steadily tightening noose of catch-and-release regulations arrived, most guys went to the Gold to kill a couple of fish. Cancelled punch cards were the metric of success. When the realities of declining stocks became obvious and the front half of the annual steelhead run timing curve was disappearing, full-on catch-and-release rules descended (see Appendix for details). Within three years the Gold and numerous other rivers saw half their angling traffic disappear as the traditionalists refused to participate. Those who remained tended to be less interested in harvesting fish and were perfectly happy to experience reduced competition. All else being equal, it doubled their catch rate. Success thereafter was all about how many fish any of us could release on a given day or trip. With no bodies or punch cards as proof, the stories that pervaded the angling community were all about how many fish had been "hit" by top guns. The numbers took on new dimensions. Thank heaven the Internet was still years in the future. It was not uncommon to hear reports of a couple of hot shots hitting as many as 50 fish on a day or two of fishing on prime water conditions. I was always a skeptic of what constituted "hitting a fish," preferring instead to measure results in terms of how many fish I had to bring to hand to remove a hook in order to release it. There were big days, though.

If catch success was the barometer, the best of the best of the Gold winter run steelheading I experienced occurred in the mid-1980s. There were days that stood out in the 1970s, to be sure, but the numbers and consistency that emerged in the mid-1980s were different. Several factors were at play, none of which was obvious in the moment. Looking back, however, that timing was one generation removed from the implementation of catch-and-release regulations.

It was also a period when ocean survival of steelhead smolts, coast-wide, was at levels nowhere near seen since. Add to that the markedly reduced angler effort precipitated by the catch-and-release regulations and the stage was set for some unprecedented fishing results. Then there was the fact that I had the only jet boat the Gold River had seen to that point. Independent of those other three factors, the boat access alone easily tripled the number of fish I could molest in a day. I should add, too, that I was as dedicated a gear fishermen as ever fished that water. If I didn't fully appreciate going in how deadly bait fishing was, it was confirmed beyond the slightest lingering doubt by the time the mid-1980s had passed.

So, what did the numbers look like? Detailed records kept during any given fishing day and diaries completed afterward are an invaluable resource to draw from 30 and 40 years later. Without them I would never have been able to recall accurately just how many days of exceptional catching I partook of. Careful review of those old records allows a perspective anglers of the present may struggle to believe.

Leading into the winter of 1982–83 I had organized a radio telemetry program to examine questions around the vulnerability of the early run timing (December and January) winter steelhead relative to those that followed in March and April. Yes, there were enough December steelhead still present to ensure the prescribed number, sex and condition of fish could be had. In company with one other staff member, weekly visits to the Gold were made to capture and radio tag suitable specimens and track every fish tagged to that point in time. Anchor tags were inserted in fish caught but not needed for the telemetry program. A jet-equipped sled was the ticket to both tagging and tracking through the lower river. During that winter two of us had days of 23, 24 and 33 fish landed, not to mention numerous other days of 10–20 fish landed. Those numbers would have been significantly higher had we been just fishing as opposed to compiling all the data that went along with each fish captured and followed with the radio receiver.

During the following winter, 1983–84, when there was no professional time involved, my records indicate the most lucrative days as 22, 41, 35 and 21 fish landed. The winter of 1984–85 followed the same pattern, with two 20-fish days, two 21-fish days and a 27-fish day. I fished alone on three of those nine days. The other six never involved more than two of us. Curious, I added up all my Gold River numbers from the 1984–85 season. It worked out that I landed an average of slightly better than 11 steelhead per day, regardless of whether that day consisted of an hour or a daylight to dark session. The fishing seemed to be holding up through the winter of 1985–86. The best of it that season was 42 fish landed for two rods on March 11. Competition and expectations were heightened by that time, though. Good fishing never stays secret. A new wave of anglers, interested in numbers not punch cards, and a different generation of guides looking to capitalize on the free lunch descended on the river. Whereas angler numbers never returned to their pre-catch-and-release era level, they brought boats that more than offset the difference, especially from a fish perspective.

There was a sidebar to that 1982–83 radio telemetry program worthy of record. During the construction of the road between tidewater and the logging camp and, eventually, the village of Gold River 8 miles upstream, there was a considerable amount of rock drilling and blasting required to carve a path along the canyon walls above the lower reaches of the river. At a point exactly 3.3 miles upstream from the tidal boundary the river was already constricted in a deep canyon but, in observing that point long after the road was a fait accompli, I saw much evidence that the constriction had been worsened by rock side cast from the road work above. Such were the road building standards of the day. The resulting cascade didn't stop steelhead but it definitely put a pause on their upstream movement. The only way of accessing that point on the river was via jet boat from downstream. On January 5, 1983, I ran our radio tracking boat to the head of the canyon pool immediately downstream, climbed the chute spilling out of the pool at the base of the obstruction pool immediately upstream and tucked the sled into the only place possible, a convenient boat-sized pocket just off to the side of the plunge pool spill. I dubbed that piece of water the Circus. The name stuck. On that inaugural visit to a piece of water that I'm confident was being fished for the first time ever, we landed ten fish. When jet boat traffic on the lower Gold increased to the point of conflict short years later, the Circus was closed to fishing. Power boats were banned from the river at about the same time. As the original river-running jet boat operator on Vancouver Island I knew full well what influence they brought to bear. Eliminating their use on the Gold was the right decision. Full credit goes to the Campbell River Branch of the Steelhead Society of BC, led at the time by Dave Hadden, for making it happen.

In 1986 I departed Vancouver Island to take on the senior fisheries management position for the provincial government's Skeena Region headquartered in Smithers. Near the end of 1999 I returned, well aware it was not the place I had

The inaugural visit to the "Circus," Gold River, January 5, 1983.

left but morbidly anxious to revisit former haunts the first winter back. I soon discovered what can happen in 14 years. It wasn't just the fact that the estimates of ocean survival rates for seaward migrating smolts had plummeted from an average of 15 per cent in the mid-1980s to 2 per cent or less over those years. That doesn't hit one in the face. What did, however, was the state of the habitat expected to produce those smolts and the condition of the river that greeted returning adults. I promised myself I would one day try to explain and illustrate the demise.

As was the case for the Englishman, the beginning of the transformation of the Gold's habitat from what I knew in the 1970s to what it is today can be traced to a single event. The date was November 13, 1975. An office colleague who toiled in the wildlife end of our business invited me to join him for a long weekend of deer hunting in the Nimpkish Valley. The wolves had not yet exacted their toll on the abundance of blacktails in that valley, and the promise of a winter supply of premium venison was all the convincing I needed. It rained hard on November 11 and harder on the 12th. Halfway through that second day we knew we were into a classic rain-on-snow event that was turning every drainage path into a torrent. We turned tail for Gold River just in time to avoid being stranded by washed-out bridges and culverts. A grader operator desperately trying to keep a section of road from being completely gutted by rising floodwaters cabled up my truck and towed us through the worst spot. We made it to Gold River just after dark and hunkered down in the canopy of my old Ford pickup beside the baseball diamond at the confluence of the Heber. The rain continued all night. At daylight I jumped out of the back of the truck and landed in water. It was mind boggling to realize the river had risen to such a height. My partner and I immediately scrambled for the cab of the truck to beat a retreat to higher ground. Nerves settled, my first thought was to grab my camera and record what I was certain was an unprecedented event. The next few hours were dedicated to precisely that.

It turned out that the flow records compiled by the Water Survey of Canada, the federal government agency responsible for such things, didn't bear out my conviction that I was witnessing history on November 13, 1975. According to them there was a nearly identical event on November 19, 1962, and an even larger one on November 15, 2006. I pointed out that the 2006 event did not see a wall of water overtop the pulp mill road to the extent that whole trees were being whipped across it into giant back eddies that formed from the force of the water hitting the rock face on the side of the road opposite the river. I offered pictures as proof and suggested they might want to revisit their flow estimates. I also pointed out that the ballpark was nowhere near flooded in 2006. I can't speak to 1962 but neither could they.

Details aside, the devastation that occurred in that 1975 flood was readily apparent when the river returned to a normal winter stage several days later. Egan's Run was obliterated by the outwash from the Muchalat system. Most of that came from the first tributary to the Muchalat River downstream from the lake outlet. That watershed and the Oktwanch at the head of the lake were literally skinned by the loggers. That roommate of the gatekeeper I conversed with a decade earlier wasn't going to be catching nine steelhead in the Oktwanch any more, before or after Christmas.

The edges and tailouts of all the major runs everywhere downstream from the Muchalat confluence were layered with unbelievable amounts of freshly deposited sand and gravel. The large pool at the Big Bend now had a gravel bar in its middle. The alder tree on river left near the tailout of the pool had always stood as my stream gauge. It disappeared that day. A run I referred to as Merganser, in honour of one of those fish-catching machines that nested in the hollow crotch of a twin-stem cedar three years running, was my first choice when the canyon waters below were too high. It couldn't be fished properly from the roadside so I used to cross slightly upstream, hide my raft in the bush and pick my way along the river's edge to its top corner. That run alone could consume half a day. The flood reduced its length by half and filled the far edge and tailout, rendering the trip across pointless. Timmons' or Clark's lost its inside edge and a good bit of its overall length. Its reputation as the money run for winter steelhead was about to fade. Downstream the devastation continued. Where the canyon reach ended about two miles above tide and the river could finally spread and lose some of its flood velocity, the deposit of finer material was immense. Areas along river left in the first mile above tide had once been a standby for intercepting incoming fish as they moved along that edge in water too deep to allow even one step from shore. The old growth cedars that held that bank together for centuries were cut long ago, as evidenced by the spring board notches on their flared stumps. Nonetheless, those stumps still held that riverbank together and protected the steelhead travel corridor for at least 30 years after the original cut. They disappeared that day. In their place was a sandy beach. Just downstream, on one of my all-time favourite runs where my friend Ted Harding and I enjoyed some of our most memorable fishing days, the formerly perfect cobble-lined bottom was smoothed and flattened by substrates too fine to appeal to steelhead.

As with the Englishman, the Gold River habitat suitable for fish and fishing just kept deteriorating. In terms of the capacity of the watershed to buffer the rain-on-snow events so common to west coast Vancouver Island drainages, the tipping point had been reached as of the event I witnessed. Too much timber yarded off too many steep slopes too fast, too many miles of road constructed from highly erodible material and with drainage structures nowhere near adequate to handle predictable volumes of water. The same rain that upland slopes would absorb and release slowly in years gone by now scours the hills and flushes through the system in ever sharper spikes on the hydrographs. Windows of optimal fishing flows that would last for three or four days following a freshet have been reduced to hours. Where once the Gold's fishing habitat remained substantially unaltered from year to year, it now follows a continually shifting design wreaked by a forest industry exempt from responsibility or accountability. The pulp mill is no longer and most of the logging and sawmilling a distant memory. With them went half the town's resource-dependent residents and businesses in a movie played over and over in British Columbia. It only took about 40 years, not more than eight steelhead generations, to bring a Vancouver Island treasure to its knees. Pictures tell it better than words.

The most instructive feature of the catastrophic changes to the Gold stems from some of my first serious forays in the early 1970s. By then I had discovered the benefits of a boat. A 9½ ft. Zodiac Cadet was my craft of the day. Some days I used a 9.8 HP prop-driven Mercury outboard along with it. What is incomprehensible when looking at the Gold River today is how a prop motor could ever have negotiated its lower reaches. The boat launch sandwiched between the original Indian Reserve community and the pulp mill was the best takeoff point to access that lower river. From there, on a decent fishing flow, I could run all the way to what was then called the Lake Pool, about three miles upstream and just downstream from the run I dubbed the Circus. Those miles bracketed the most generous winter steelhead water and, if well timed, summer steelhead water as well. Such was the depth of the water through all the canyon pool tailouts and the riffles. In the early summer of 2013, on the same volume of water once considered prime for encountering steelhead, I revisited my old haunts on an experimental mission with a jet pump–equipped sled. I dared not even try the first significant riffle scarcely a half mile beyond tidal influence. There was simply not enough depth to avoid collision with the bottom whose configuration had been altered by almost four decades worth of material washed from logged headwater slopes.

The author navigating the lower Gold River in his prop motor-driven Zodiac in June 1974.

The first four photographs on page 68 were taken from the same spot over a period of 39 years. The top photo shows my friend Ted Harding fishing one of our most prolific summer steelhead runs on May 18, 1974. The flood waters of November 13, 1975 left the bank behind where Ted stood beginning to erode and the cobble substrate through the body of the run being replaced by ever smaller material. By 1983 the bank on river right was showing progressively more erosion and deposition of fine material along its edge. The formerly productive holding water midstream lacked the velocity it once provided, and the accumulation of finer materials midstream was clearly evident. The first visit to the same piece of river following a 13-year sojourn in Smithers was sobering. The date was November 2003. A strong wind and a very high tide that day partially backwatered the run and obscured its substrate. Nonetheless there had obviously been a tremendous accumulation of outwash from higher in the watershed all along river right. Periodic visits in subsequent years revealed constant redefinition of the channel and beach all through the area.

Time series photos from the same vantage point but looking downstream rather than across gave more perspective on the changes that occurred. Looking back, I wish I had taken more regular photos of that downstream view from the rock bluff. A decade or more between the shots does not come close to documenting just how many times the channel shifted.

Habitat degradation is a big part of the Gold River story, to be sure. The river's capacity to produce the smolt crops of yesteryear is no longer. It looks and feels sterile. Fewer smolts sent to an ocean that doesn't welcome them as it once did compounds the problem. Together those factors would explain an overall reduced abundance but not the virtual disappearance of the front end of the winter steelhead run. Angling couldn't possibly be a factor – or could it? Before pronouncing on that, a bit more rationalization of observations and experiences is in order.

Some will be quick to note that my case for habitat degradation pursuant to the November 13, 1975, flood event being responsible for diminished fish returns doesn't fit with the catch success many of us experienced eight to ten years later. I've struggled with explaining that myself. The retrospective view confirms that the smolt-to-adult survival rates among steelhead stocks all along the Pacific Coast in the mid-1980s were sufficiently high to offset the habitat destruction influence or at least disguise what should have been obvious. In the thick of those years the prevailing professional opinion was that ocean survival was unlikely to persist at those levels but that the good fishing was enough to subdue any serious consideration of the longer term. After all, we had catch-and-release as our salvation. By the early 1990s the smolt survival pendulum swung wildly in the opposite direction, leaving anglers and professionals equally distressed. The double jeopardy of diminished capacity of freshwater habitat to produce smolts and survival rates at sea that were as low as ever measured gave rise to adult returns and catch rates that were polar opposites of those of the mid-1980s. No sign of any significant or sustained reversal of those circumstances is evident today.

Now, back to the business of angling influence. I'll not try to tell the world there was any guarantee a proficient rod would find newly arrived winter steelhead in November, but I will say it wasn't a rare occurrence in the 1970s. The mid-1960s report of those nine fish the Oktwanch gatekeeper's logging camp friend had caught by Christmas implies there were good numbers early on, though. Tim Timmons, the guide connected to the Gold River Inn during the pulp mill start-up years, boasted November steelhead on several occasions according to newspaper columnist Lee Straight. December was much more predictable in my earliest years. It was a major disappointment if a weekend trip from former residences in either Victoria or Nanaimo to Gold River any time after the first of that month didn't produce multiple fish. And, as noted earlier, the week between Christmas and New Year's was busy and bloody. It took a few years to realize that the front end loading of the annual harvest was likely taking a toll of the most valuable fish the river offered, those that came the earliest and stayed the longest. That precipitated catch-and-release regulations near the end of the 1979 season and, within three years, for the entire season. The objective was to prevent further erosion of the front half of the run timing curve and, hopefully, assist it to rebuild. It didn't happen. Any November and December run timing genes that may have survived the harvest era were loved to death by the catch-and-release artists, myself included, who believed there would never be any harm done by catching the same fish over and over. That was then. Today I'm firmly of the opinion that we anglers are largely responsible for the fact that most of what remains of the Gold's winter steelhead originate from parental stock programmed to return after the peak of both the harvest and catch-and-release fisheries characteristic of that stream in the first 20–25 years following public access. The selective advantage afforded the relatively lightly fished March and April fish has resulted in that component representing a far greater proportion of today's run timing curve. That outcome is anything but unique to the Gold.

There is one last feature of the Gold steelhead fishery I should mention. When the hatchery steelhead programs were ramping up in the late 1970s and into the 1980s, it was not uncommon to encounter adipose-clipped steelhead in the lower Gold River during the winter fishery. I once caught three in a day. The fish were always on the small side compared to typical Gold River winter runs and often appeared to have been in fresh water for some time. It is possible at least some of those fish were summer run stock. The recovery of kelts on some occasions was evidence the fish were not necessarily temporary residents. Coded wire tag analysis revealed that the clipped fish were always strays from Robertson Creek Hatchery on the Stamp/Somass system, fully 165 water miles distant. We joked that the fish turned left one pulp mill too soon on their journey south from their ocean pasture. The pulp mill influence may not have been insignificant given that strays from Robertson Creek were never reported from other rivers between the Gold and the Stamp/Somass.

A chilling feature of the influence of strays surfaced many years later when DNA analysis of chinook from the Gold system revealed that those fish were indistinguishable from Robertson Creek Hatchery chinook. Hatchery volunteers untrained in the nuances of brood stock collection and spawning protocols failed to recognize the steady decline and eventual disappearance of the older, larger, Muchalat-origin chinook the area was once famous for. That and the presence of known strays from Robertson Creek inspired the geneticists to look deeper. Their results confirmed the Muchalat-origin chinook had been extirpated.

Time series photographs of the same run on the lower Gold River between 1974 and 2013.

Comparative views of one other section of the Gold River in 1973 and again in 2012.

Comparative views of the same rocks on the lower Gold River in 1976 and 2012.

Comparative views of the lower Gold River at the tidal boundary on three different occasions between 1974 and 2013.

An illustration of the fishing available on the lowermost reaches of the Gold River before the catastrophic flood event of November 13, 1975.

Opposite page and below: Comparative views of another section of the lower Gold River in 1974 and 2012.

The Heber River as it appeared near the peak of the November 13, 1975 flood event and again, at an average winter flow on March 21, 2014.

Comparative photos taken at the same location on the highway between Gold River townsite and the pulp mill on November 13, 1975 and in the spring of 1976.

Comparative views of the "Number 2 Bridge" over Gold River near the peak
of the November 13, 1975 flood and again in spring 1976.

6

KISPIOX SOUTH

It would be hard to find an experienced steelhead angler who is unaware of the Kispiox River's storied history of producing the largest steelhead in the world. Devotees of other Skeena tributaries, namely the Babine and Sustut, will argue their fish are as large. They aren't. Both produce large steelhead but not as large as Kispiox fish and not with the same frequency. Those are the fabled interior summer steelhead that share, in name only, an affiliation with their coastal steelhead counterparts. Fish that top the 30 lb. mark and, rarely, even 40 lb. are obviously not unknown among select interior stocks but virtually never occur among their winter cousins. Shave 10 lb. off the high end of the interior summer steelhead weight spectrum and you'll have the rough equivalent for the largest winter fish. Very few streams produce such animals, and the only one I have ever known to do so with any consistency is Vancouver Island's Salmon River. The Yakoun River on Haida Gwaii may be a parallel, although I do not have enough personal experience there to speak definitively on it.[11] After a decade of thorough investigation of the Salmon I came to appreciate that I would see a higher proportion of 20 lb. steelhead among fish I handled there than on any other river I ever fished, Skeena tributaries included – and I fished all the big ones. Creel survey records from several Skeena tributaries, including the Kispiox, confirmed that observation. More about the fish later, but not before some background.

I missed the best of the Salmon River by three-quarters of a century, if not more. By the mid-1970s when I began my professional association, the valley bottom was long since stripped of every merchantable stem. Huge, decaying conifer stumps with springboard notches stood testament to the forest giants with massive interconnected root systems that once held the valley floor together. I marvelled at the spacing of those stumps and how the river unquestionably ran a stable course in order for trees along its banks to have achieved such age and size. By the time I arrived the roots that were the glue of the riparian zones were long rotted away, leaving tunnels in all the exposed cutbanks that had been created by decades of post-logging freshets. There was no hope of the floodplain being stabilized by the puny, overstocked hemlock and fir trees or the jungle of alders and willow that carpeted the riparian zones. Acres of stream banks washed away every time the rains came. With them went the next generation of forest. I don't recall a single grove of old growth timber anywhere along the Salmon River channel reaches accessible to steelhead at that time. The river channel showed all the telltale scars as well. It was a magnified edition of the Englishman, the only difference being the Salmon had a head start.

Perhaps the most telling comment I can relate regarding the Salmon was made by a respected professional colleague who was the senior fisheries biologist for MacMillan Bloedel, the forest licence holder for the Salmon River watershed at the time. I distinctly remember him telling me the inside word among company officials was that their Kelsey Bay Division (the one that feasted on the Salmon) had taken more timber out of the Salmon watershed than the combined total the company had removed from all its other watersheds on Vancouver Island. The magnitude of that is understood when one considers that they held the forest licences from a couple of watersheds north of the Salmon all the way to just north of Victoria plus a fair number of watersheds on the west coast of the southern part of the Island.

A bit of the flavour of what had once been there was covered in another chapter in Lee Richardson's book referenced earlier. Richardson's association would seem to have begun not long after the end of the Second World War, although specific dates were never given. He spoke of driving 50 miles to the Salmon from Campbell River on a preposterous single lane road barely negotiable by the station wagon his party had at their disposal. His description of the times, not unlike that for the lower Gold River, is worth repeating:

> Before the loggers harvested the timber at the turn of the century, the valley of the Salmon sustained a forest of Douglas firs without peer in the entire Northwest, many of which were old when America was discovered.... Today all that remains as a reminder of the glory that once was are some enormous moss-covered stumps hidden in the alder and hemlock jungles.... Indeed, that era is gone now, a page in our lives than can never be rewritten.[12]

Richardson's mission was fish, specifically cutthroat trout. He spoke of the riffles in the lower reaches of the river, near Sayward, as "being in a perpetual state of upheaval due to the highly agitated multitude of fish." Huge chinook salmon cruised about in full spawning ritual with cutthroat, Dolly Varden and juvenile steelhead spread throughout. Too many prized cutthroat flies were lost to chinook he tried his best to avoid hooking. Tidal waters below were described as chock full of big burly coho eager to hammer bucktail flies. "This was coho fishing at its glorious best – with the possible exception of fishing on the Queen Charlotte's Tlell." Such was the state of the Salmon a lifetime ago.

To understand what remained of the Salmon River and its steelhead as of the latter stages of the 20th century one must know something more of the nature of the river by that time. It is the third largest watershed on Vancouver Island, behind only the Nimpkish immediately north and the Campbell, which borders it to the south. Only a small fraction of the Campbell was ever accessible to anadromous fish, though. Unfortunately, Salmon River steelhead did not have access to most of the best habitat in their home river either. Two barriers blocked their path. One consisted of a plug in the only section of the river that would qualify as a canyon. It was located about 28 miles upstream from tidewater and looked to have been a consequence of logging railroad construction and a crossing that once existed at that point. At least one immense rock was lodged between the canyon walls, and all the outwash and debris that accompanies logging had accumulated behind it, creating a vertical drop too great for any fish to ascend. Next was a water diversion dam courtesy of the power production giant, BC Hydro.

At roughly the same time Roderick Haig-Brown was campaigning to save the upper Campbell River system from the inevitable consequences of hydro dams, a diversion dam was being constructed on the Salmon about eight miles upstream from the migration blockage. That 1958 dam directed most of the Salmon's flow, at critically important times for fish, south to the Campbell reservoir system to augment the power production available from the generation facilities there. It will never be known whether the mini-canyon migration barrier prevented steelhead (and other species) from utilizing the river reaches beyond before the loggers arrived. Given the nature of the barrier and the extent of logging that occurred in earlier years, it doesn't seem an unreasonable assumption that it did. Regardless, the dam was never intended to pass any steelhead that did reach that point.

The fisheries authorities, myself included, undertook an array of projects intended to ameliorate the damage done by both logging and BC Hydro. First was removal of the migration barrier in the canyon downstream from the diversion. Drilling and blasting of logs and rocks followed by annual freshets eventually wore away the blockage giving fish access to the reaches upstream to the base of the diversion dam. That was accomplished by about 1980. Coincidentally, engineering staff developed a screen that was inserted into the canal leading away from the dam site. The screen was intended to trap emigrating steelhead smolts and send them back into the Salmon instead of down the concrete diversion flume and eventually into the Campbell River reservoir system, where they would be landlocked. The iterations of the smolt trap in the diversion canal and the process of redirecting them back into the mainstem Salmon River became a growth industry. As always, the lion's share of the resources dedicated to "fixing" the steelhead problems have never come from the loggers or power producers.

The other major initiative for the Salmon was colonization of the watershed upstream from the dam with steelhead fry originating from parents collected below. In concert with fertilization of the areas where fry were being released, the hope was that natural production of steelhead in subsequent generations would be jump started. Utilizing available steelhead and salmon–producing habitat upstream from the diversion dam was a readily saleable initiative if the problem of loss of emigrating smolts could be overcome. Most of my experience with Salmon River steelhead stems from organizing the capture of the brood stock required to fulfill the fry stocking program commitment. It was an undertaking I look back on with mixed emotions. I've since taken the position that our sense of omnipotence around helping steelhead do what they do best was ill-founded. They would have been better off without our intervention in their reproductive distribution and output. Some remarkable experiences never duplicated at any other time or place in my steelhead career stem from those years when we believed fervently we could help, though.

I should say here that 30 years later there are fish to the base of the diversion dam. How many of those may have resulted from the fry stocking efforts and how many from natural colonization of the eight miles of river between the former migration barrier and the diversion dam is an unanswerable question. And, not surprisingly, the politics and business of preventing smolts from being siphoned into the Campbell River reservoir system and getting adults past the dam, both upstream and downstream, continues.[13]

Now, about the fish and fishing. With the aftermath of logging abundantly evident, it was always an adventure to move around on the Salmon. The valley bottom was all but impenetrable by foot except when the latest shift in the stream channel brought it close to an old logging access spur. The river itself was full of stumps and debris that forced continual braiding and redirection of the channel. It could be a very different river from freshet to freshet, let alone between years. Finding fish was a relentless exercise in rediscovery.

Comparative views of the righwt bank of the Salmon River at the
Memekay River confluence in March 1980 and January 1981.

Comparative aerial views of the Salmon/Memekay confluence area
on March 27, 1986 and again on October 24, 2002.

In 1980, when the steelhead brood stock initiative was taking form, the immediate task was to familiarize ourselves with the river and how best to allocate our time. A jet boat trip from Kelsey Bay near the mouth of the river to as far upstream as safe passage would allow topped the agenda. The first ten miles of river were quickly revealed as steelhead angling purgatory. It was an utter waste of time to bother with those slow, lake-like, sand and silt–lined, debris-choked areas if the intention was to find any accumulation of fish. The odd one might be encountered in transit, but there was no typical holding water and almost nothing that appealed to a fisherman. Once beyond, the river gradient started to increase such that the occasional piece sported depth and velocity that gave some promise of harbouring fish. At a point known as Pallan's Bridge (long since washed away), about 17 miles from tidewater, an old access road between the highway a mile or so distant and the river proved to be the best place to launch a boat. Experienced eyes determined that the river from there upstream to the Big Tree Creek confluence at mile 20.5 and, especially from there to the Memekay confluence, slightly more than a mile farther, should be the focus of attention.

At a reasonable fishing level the river was readily navigable from Pallan's to Big Tree. From there to Memekay was much more challenging due to channel braiding and the subtraction of Big Tree's contribution to river volume. Beyond the Memekay the flow was halved, making any further access by jet boat more than slightly hazardous. No matter – most of the Salmon's steelhead did not go beyond Memekay, at least not at that time.

Jet boat adventures on the Salmon were many. Just getting a boat in and out of the river at our chosen launch spot at Pallan's was a major chore. One of our most memorable trips involved a small-scale logging operation just to cut our way through flood-deposited trees all along the road between the highway and the launch site. I clearly recall that the tool kit we had on the floor in the cab of our pickup was floating as we made our way through the lowest spot on the road that day. Once in the river and headed upstream we found the channel blocked in several locations between Big Tree and Memekay. Out came the chainsaw once more. My partner that day, Brian Clozza, was an experienced faller and I an experienced boat operator. With me at the helm feathering the throttle to maintain position and Brian jumping on and off the bow of the 16 ft. Smokercraft to beaver away at log after log with a 40 in. bar Stihl chainsaw, it made for violation of every safety standard imaginable. No one ever accused us of not being keen.

There were other jet boat tales to tell. At the conclusion of a long couple of days of collecting and transporting fish and safely transferring them
to the oxygenated tank in the back of our pickup, my partner of the day, Lew
Carswell, and I proceeded to load up the boat and make for the Quinsam River
Hatchery at Campbell River to deliver the cargo. We had a heavy load and a
two-wheel drive truck to get our boat onto a trailer and out of the river. The
only way to make it happen was to run the truck across a bit of a side channel
at the Pallan's site to reach a more solid and favourably sloped gravel beach
immediately downstream. Mission accomplished. The boat was loaded and we
got up a bit of a head of steam to make it across the side channel and up onto
what remained of the road out to the highway. Well, the head of steam produced a head of steam. It seems we managed to get about one cupful of water
through the carburetor and seize the engine just as the truck cleared the river.
With darkness approaching, a valuable payload of live fish, miles to go for help,
and no communication, the only option we had was to relaunch the boat and
make for the gas station at White River Court just up from the confluence of
the White River and the Salmon. We figured it to be nine or ten miles and half
an hour distant. Luckily the boat trailer didn't make it far enough out of the
water to prevent relaunch.

Clearing logs off the flooded road between the North Island Highway and the
rudimentary boat launch at the old Pallan's Bridge sit on the Salmon River on
January 22, 1981.

Clearing a navigation channel through the log-infested Salmon River between the Big Tree Creek confluence and the Memekay River confluence on January 23, 1981.

Off we went downstream through water we hadn't visited for more than a year. The plan felt good until we rounded a bend about halfway along and found the channel completely blocked by a pair of large spruce trees. One was at water level with a film of flow over it. The other paralleled it a couple of boat lengths downstream but about four feet above the surface. Thankfully the river was as placid as it gets so we didn't have to fight current speed or turbulence. Risk-averse Lew was a hard sell, but I managed to convince him we could nudge the bow of the boat up onto the first log and then power it far enough up to use it as a fulcrum. Then, if we killed the motor and shifted our weight to the front, we could teeter totter the boat enough to eventually rock our way over to the downstream side. It worked, but now we still had to get past the second log. With the motor in reverse to keep the boat speed ever so slightly less than the current speed, we inched the boat underneath until it came up against the highest point, the steering wheel on the centre console. It jammed against the underside of the log and began to push the boat under. Just when we thought we were trapped, the pressure on the steering wheel bent it enough to burp the boat loose on the downstream side of the tree. Another bullet dodged, we powered up and made for White River Court. All that remained was to convince Jack, the gas station proprietor, to grab his tow truck, ferry us back to Pallan's, retrieve the truck and its precious cargo of brood stock and tow us all to Quinsam Hatchery. It was a long day.

Retrieving the jet boat and steelhead brood stock transport truck on the Salmon River near the old Pallan's Bridge site in late winter 1984.

One last jet boat story to illustrate the nature of the river and the obstacles that had to be dealt with to achieve the objectives around securing the prescribed number of male and female steelhead for the colonization efforts in the upper Salmon River. This time my partner was Maurice Lirette, a keener if ever there was. The infamous day was February 22, 1984. It was our first visit to the Salmon that winter, and our intention was to check out the access limitations another season's worth of freshets had delivered to that debris-laden flood plain. Getting to the launch point at Pallan's was uneventful, as was the trip upstream as far as the Big Tree confluence. From there to Memekay was a maze of braided channels with passage often constrained by stumps, log jams and sweepers. Mental notes were made regarding the precise path to take on the downstream leg. Those familiar with jet boats in small water understand the difference in the degree of difficulty between upstream and downstream travel.

Late in the day, on the return trip between Memekay and Big Tree, we had to negotiate a dogleg that was bounded on river right by a log jam with a sweeper extending halfway across the desired path. River left was the inside bend of the dogleg and consisted of a gravel bar that reached far enough into the channel to make the passageway about the same width as the boat. The combination of direction of flow, velocity and channel dimensions meant I had to run the boat straight at the logs and then make a hairpin left, keeping the motor in the deeper water near the logs on the right while skating the bow across the shallows at the toe of the bar on river left. What I couldn't avoid entirely was the sweeper. Just as I made the sharp left, the stern of the boat skated slightly to the right and caught the submerged end of the sweeper. It took only seconds for the current to push the boat steadily along it until the jet pump intake was nicely out of the water, rendering us helpless. At that point we knew the force of the current against the tilted hull was going to force the port gunwale under, flip the boat and paste it against the log jam. Start to finish was not more than 90 seconds but it felt like a feature length movie in ultra-slow motion. We managed to throw everything mobile onto the logs and scramble out ourselves before the river eliminated the option. It was a long and arduous hike through the tangled mess of valley bottom jungle to get to our truck and retreat to Campbell River to figure out what to do next.

Inspired by strong coffee the following morning, we developed a plan to retrieve the drowned boat. Our first stop was the helicopter base and the people we had relied on multiple times for various fish-related undertakings, most notably the steelhead fry releases in the upper Salmon River. We convinced them it was worth trying to lift the boat out of the logs by connecting a longline to its bow eye. Less than an hour later we were on site carrying out the plan. It worked to a tee. The boat came free surprisingly easily. In an impressive display of skill the pilot then guided it across and gently lowered it in the shallows opposite the sweeper that claimed it. Elapsed time, not more than ten minutes. After connecting a fresh battery, pushing double mix gas through the system by manually turning the flywheel and removing, drying and replacing spark plugs two or three times, we were amazed to have the motor fire up as soon as the key was turned. The boat and motor were well washed but otherwise unscathed. Both were still on duty when I last saw them two years later.

I said earlier the Salmon River's steelhead were in a different class in terms of the proportion of them that exceeded the popular benchmark of 20 lb. To support that remark I revisited my personal archives and uncovered photos I had taken of 17 steelhead in excess of 20 lb. Every one of those fish was caught either by me or my partner of the day. The time span covered by the photos was five seasons. There was probably that many again that didn't get photographed for one reason or another, and there were twice that number that fell in the range from 18 to just under 20 lb. Our best days on the Salmon rarely matched an average day on the Gold at the same time, but I can count on a hand and a half the number of 20+ lb. steelhead I witnessed on the latter over a much greater sample size and a far longer time period. That aside, I'd be remiss if I

Retrieving the jet boat from the Salmon River between Big Tree Creek and the Memekay River confluence on February 22, 1984 after the sweeper log incident the day before.

didn't acknowledge the behemoth Gold River steelhead caught by another of
Campbell River's iconic anglers of yesteryear, Bruce Gerhart (RIP), on January
27, 1994. Bruce's fish weighed 30 lb. 11 oz. and stands as the largest Vancouver
Island steelhead I have ever known. It is prominently and proudly displayed as
a sign of its time at River Sportsman in Campbell River. I think I knew Bruce
well enough to appreciate that he would have had difficulty dispatching such
a magnificent fish even though it was entirely legal. He was one of the most
forward-looking steelhead anglers of his time and a strong advocate of catch-
and-release long before it became mandatory and accepted.

More evidence of the frequency of large Salmon River steelhead stems
from old accounts of annual derbies referenced by long-time outdoors writer
for the Victoria *Times Colonist* newspaper, Alec Merriman (RIP). In 1967 he
reported that Ross Spiers was the winner of the contest for the aggregate
weight of three fish over 20 lb. Spiers's total was 63.4 lb., His friend Bill Clark's
three came second at 61.5 lb.

The trick to finding the Salmon's big ones was very straightforward.
Locate decent holding water and dredge a piece of good bait through it. In
some years the flood waters would leave behind the odd piece of water in the
Big Tree and Memekay areas that had enough gradient to provide something
larger than gravel for substrate and a bit of turbulence for overhead cover.
Elsewhere, though, almost all the holding water was characterized by log jams
and stumps scattered through low-gradient, low-velocity areas. Snorkel observations confirmed that in spades. Our objective was always to maximize the
number of fish caught per unit effort, so anything less than the most effective
gear known was never on the agenda. That meant bait. Conventional "hardware" just didn't hold a candle to bait, even at higher flow stages when the two
terminal gear choices sometimes produced equally well in other places. Only
bait would elicit the addictive response that would tease fish out from under
the Salmon's logs and stumps.

Salmon River steelhead made for as many memorable moments as did
its jet boating adventures. There was the day when the logs claimed seven
of seven heavy fish hooked. There was another where two successive casts
brought fish to hand without either being impaled by a hook. On the first
the leader wrapped around its snout and a split shot pinched tightly enough
behind a maxillary to seal its mouth and allow enough pressure to be applied
to bring it ashore. There was no hook to be found when I unwrapped the leader.
The very next cast found another fish that tore off downstream, daring us

to chase it with the jet boat. That was the one and only such incident on the Salmon. Hundreds of yards downstream the too-regular throbbing of the rod tip signalled that it was hooked in or near the tail. When we finally caught up we found the leader had looped around the caudal peduncle and hooked over itself, tightly securing the fish. They fight hard under those circumstances. Neither of those fish broke 20 lb. but they weren't far off.

The day when a seal surfaced in the middle of a run about 18 miles upstream from tidewater was another of note. It was mid-March in 1984. We had just pulled in to the first piece of water we intended to fish that day. I took the back half of the run, my partner the top. As I waded in and was about to swing away with my first cast, a shiny black head emerged 20 feet out. I shouted to Maurice, "Did you see that?" He did. The only way that animal could have reached that point was to have moved upriver on the latest high water. That had been 8–10 days earlier, and the river was now down to the point where our seal was trapped by several hundred yards of shallows barely negotiable by jet boat immediately downstream. Fishing was pointless so we decided we should get even with the seal. With my accomplice on the throttle and me hanging over the bow armed with an oar, we herded the seal downstream until it was running out of water. It disappeared underneath the bow just as I was about to try for a home run swing. It couldn't escape the shallows, though, so we had it sandwiched between the hull and the bottom. The shoe of the jet hit the seal's head and bounced hard skyward, cavitating loudly. We throttled down, looked back and saw the seal make for the logs in deeper water upstream. It disappeared into the maze never to be seen again, at least not by us. I have no idea whether that animal ever returned to its normal haunts, but it couldn't have done so without much a much greater volume of river than was on the horizon for the next many days. No doubt it dined on a few steelhead in the meantime.

Snorkel surveys undertaken periodically to locate concentrations of steelhead catchable by beach seines provided the stuff of more good stories. Wildlife tend to be less than wary of anything approaching from water. The Salmon River Valley elk were no exception. On one of those snorkel days on the reaches between Big Tree Creek and Pallan's Bridge, two of us slipped along quietly in our dry suits until we were within ten feet of the nearest of 11 elk bedded down on a comfortable-looking sandy beach. The only member of the herd that was the least bit alert bore a yellow radio telemetry collar that had obviously been applied by one of my wildlife colleagues weeks or even months earlier. Not until we stood up in the water did that elk rouse the rest of the herd and crash off into the streamside jungle. Their rapid departure stirred up the mud along the edge of the river such that we had to take a serious time out to wait for the pulse of dirty water they created to flush itself through so we could continue our survey.

On another occasion, while snorkelling down the Memekay River only a few hundred yards above its junction with the Salmon, my partner and I took opposite channels created by a small midstream island. He spooked a black-tail doe which made a hasty departure from his side of the island but stopped in the channel I was drifting down. I remained motionless and let the current carry me straight ahead. The closer I got to the deer, the more I became concerned it was not going to move. It didn't. When the collision course was set I reached out and poked the left rear leg with my right hand while keeping my left hand and arm protecting my face. The deer wheeled around and bolted for the side channel from whence it came but returned seconds later and bounced straight across the channel in front of me, disappearing into the bush in what looked to be a fit of terror. I've seen all manner of wildlife, including countless bears and moose, a half dozen wolves, a couple of bobcats, a wolverine and even a swimming cougar while snorkelling or passively boating along rivers, but that was the only time I ever touched an animal.

Now, more on those Salmon River fish. Tubes I conceived and designed myself were a big part of the process of confining and holding brood fish between capture and jet boat transport to our tank truck at day's end. There was a succession of types and styles but the best of them were zippered cylinders of flexible Hypalon fabricated by an old school British sailmaker in Nanaimo. They were about 41 in. long and 10 in. in diameter. I recall releasing two fish over concern that bending them sufficiently to get them in the tubes and keeping them in that position until they could be run back to the tank truck would do too much damage. Those fish were definitely in the mid-20s range. My friend Lew Carswell caught another of that size that ended up as the slipcover picture on a book authored by an old schoolmate of his. (The author, well known in steelhead circles, is holding the fish but gives no credit to Lew for catching it or taking the picture!)

The quest for brood stock frequently involved volunteer anglers from local communities and clubs. Whereas it was good politics and sometimes just plain fun to get together with a keen bunch of guys who shared equal passion for their favourite river, such efforts did not come without a heavy burden of coordination. Equipment had to be distributed, anglers deployed appropriately,

fish handled properly, communication maintained, tubed fish found and transported at day's end etc. That was an experience duplicated many other times
and places, but there was a particular incident on the Salmon that brought it
all home. It was mid-March 1979 and our first major effort to organize members
of the local fish and game club. We set up to have volunteers on the river for a
weekend. Nine of them showed up for day one but they only accounted for one
fish. Undeterred, we organized a repeat for the following day. Three decided to
partake. By mid-afternoon it was time to search them out and check on results.
Rounding a bend well downstream from our usual beat beyond the launch site
at Pallan's we found our last man. He was stooped over a fish still clutching a
piece of driftwood he had just used to dispatch it. The fish was later weighed
at 29.3 lb.

We forgave our friend for keeping the biggest steelhead he had ever
caught but not without hinting that we weren't all that happy about potential
brood stock ending up as wall hangings. A few days later we checked back
to see how the brood stock collection efforts were going and learned that our
man had seen fit to kill two more fish, both females and both weighed at the
local store. One was 22.7 lb, the other 18.0 lb. We also received news another
club member had killed a 25.8 lb. male at the same time. Diplomacy was put
to the test, but I managed to convey the message thanks-but-no-thanks for any
further contributions from the club.

The Salmon was never my first choice as a steelhead fishing destination, but
it is the one I most regret not having experienced before logging destroyed its
attractiveness. Imagine Lee Richardson's take on its character today. All things
considered, though, if the object is a big winter steelhead the Salmon is in a class
by itself. I'm glad to have had the opportunity for a professional association with
those unique fish. The appreciation I have for them would never have developed
in the absence of the many adventures that association involved.

Example photographs of several 20 lb.+ Salmon River steelhead encountered during the brood stock collection efforts on that river in the late 1970s and early 1980s. Cyril Webster's fish (top) at 29.3 lb. was the largest.

7

ANOTHER ONE BITES THE DUST

Vancouver Island became well known domestically and internationally as the heartland of "the war in the woods" in the decade of the 1980s. The world heard all about the fight to save the old growth timber in watersheds such as the Carmanah and Walbran and the area known as Clayoquot Sound. The Raging Grannies stood their ground in front of the logger's heavy equipment, First Nations members chained themselves to trees and equipment and protest upon protest was orchestrated from the woods to government halls. The time had come to take and make a stand. Some big trees were saved for posterity. All good stuff. What gets overlooked, though, is the fact that those places represented, not the best of what once was, but all that remained. The good valleys were eviscerated long before because they grew the most, the biggest and the most accessible timber Vancouver Island ever offered. Every fish-bearing stream in Clayoquot Sound combined never produced a fraction of what came from rivers like the Cowichan, Gold, Nimpkish, Salmon and so on and so on. I'd trade a dozen Clayoquot Sounds for one Gold River. There is a reason why Haig-Brown and Lee Richardson never wrote about fishing rivers like Walbran and Carmanah.

Before environmentalism became popular enough for politicians to take note, there was another watershed up for grabs, the Tsitika. It had escaped the fate of the other major timber-producing watersheds to the north and south because it didn't promise anywhere near the volume of timber and what was there was hard to get to. There were no roads along the jagged coast from either the south or the north that would facilitate the usual assault from tidewater upstream, and the timber value wasn't there to justify constructing them. However, when the new highway north from Campbell River cut across the top end of the Tsitika Valley on its way to Port McNeill and Port Hardy, the stage was set to get at that old growth from the upstream end.

The logging proposals for the Tsitika hit the airwaves in the early 1970s, just about the time a relatively environmentally sensitive government had dethroned its long-entrenched resource development predecessors. For the first time ever it wasn't quite as easy for the forest industry giants to have their way with fish and wildlife advocates and protectors, not the least of them Roderick Haig-Brown.

There were well-organized public hearings to gauge the strength and support for logging versus preservation positions. I attended one where I vividly recall a logging family's 8- or 9-year-old son taking the stage and telling the audience if "they" didn't allow his daddy to cut the trees he wouldn't have a house to live in and they wouldn't be able to buy food. Game on.

Looking back, I don't think any of us in the fish and wildlife business expected an outcome dramatically different than what history would predict. That should never diminish the dedication of a lot of good people who gave it their all in passionate defence of the fish and wildlife resources. Ultimately, the final nail in the Tsitika preservation coffin was a provincial election during the late stages of the debate over its future. Any semblance of environmentalism that prevailed in the early discussions on the Tsitika was soon forgotten with the ousting of the New Democrats' Dave Barrett and the return of the business-friendly Social Credit crew led by Bill Bennett. That set the stage for the Tsitika to become one more entry on the long list of watersheds irreversibly altered by access and timber removal. The preservation and protection advocates' tenacity was not entirely in vain, though. The valley's fate was a happier one for their efforts than it would have been otherwise. The lower few miles, downstream from Catherine Creek, became a park. There were a couple of ecological reserves established in the high country and another at Robson Bight, the killer whale–rubbing beach adjacent to the river's estuary. Thank heaven for those blackfish and the bona fide unique rubbing beach. Had that remarkable piece of real estate and its cetacean visitors not been there, the estuary of the Tsitika would very likely have become another log dump.

For those who think logging was prosecuted differently in the Tsitika after all the public debate about whether it should even occur and all the microscopy focused on logging plans when it was decided it would, I need to relate a bit more history. Our Fish and Wildlife office in Campbell River was the home base of the Habitat Protection staff who were responsible for monitoring the Tsitika operations. Shortly after the first fallers went to work in the upper reaches of the watershed, one of them stopped by the office and discreetly advised a staff member he needed to get out there and take a look at what was occurring. Shortly afterward charges were laid under the Fisheries Act and a long and contorted court process ensued. The very first block of timber felled following years of debate and planning aimed at a new and enlightened way of doing business – and the company ends up in court. Tune in Google Earth today and see how the Tsitika compares to any other watershed of choice.

Before the first chainsaw arrived in the Tsitika I had the opportunity to see it in its pristine state. It gave glimpses of what all of Vancouver Island's rivers must once have been. My first visit came at Easter in 1975. With the conflict around logging heating up and large gaps in the knowledge base on Tsitika steelhead, I volunteered to spend the holiday weekend to try to find evidence of winter steelhead. Summer fish were known to be present and some information on numbers and distribution had been compiled from snorkel surveys the previous summer. A helicopter dropped my partner Rick Axford and me at the confluence of Catherine Creek, the Tsitika's major tributary. We camped there and hiked the river below for the next two days. If snowstorms are the metric it was a memorable camping trip.

It didn't take long to satisfy ourselves that winter steelhead were highly unlikely to be present at anything more than a small fraction of their summer steelhead cousins' abundance. There was a major migration obstruction about one mile downstream from Catherine Creek and no tributaries between there and the estuary. That left winter steelhead not more than three miles of mainstem river habitat to sustain themselves. If they were ever present in detectable numbers April would be the time to find evidence in terms of either fish or redds. We found neither. Years later the obstruction that we observed blocking passage for anything other than summer steelhead was completely reconfigured by flood waters, leaving the river beyond accessible to even lowly pink salmon. Who knows how often such events may have occurred historically, but I'm convinced the river we visited in 1975 was the exclusive preserve of summer steelhead upstream from the point we noted as the original migration obstruction.

On one of the early trips in the quest for better information on the juvenile steelhead population strength and distribution, we enlisted the services of a marine patrol vessel piloted by one of the province's Conservation Officers based in Campbell River. The pilot ran the boat and crew to a nicely protected moorage, Growler Cove, on West Cracroft Island, just across Johnstone Strait from the Tsitika's estuary. Early each morning a helicopter picked up our crew from the float we tied up to and ferried us to our work sites along the Tsitika for the day before returning us to the cove. The impressive thing about that initial work had nothing to do with juvenile steelhead. Rather, it was the commercial fishing fleet that surrounded us at Growler Cove and how they went about their business. If ever I needed a first-hand impression of what fish were up against relative to commercial fishing harvest, that was the place to discover it.

Images of Tsitika River at and downstream of the Catherine Creek confluence when the
river was in its pristine condition prior to logging. April 1975.

The date was early July (10th–12th), 1978. Seine fishing vessels were present all through the area between our moorage in Growler Cove and Cracroft Point on the northwest tip of West Cracroft Island. What the seiners already knew, and what we learned by watching them, was the migration pattern of the mature chinook salmon they were targeting. They had it figured that the fish moved around Cracroft Point from the waters beyond in either Blackfish Sound or Baronet Pass and then southerly along the shore down into the cove. There they milled about before curling around the exit from the cove and continuing on downstream through Johnstone Strait. The pattern was so well understood by the seiners they didn't bother to fish anywhere else. They just lined up and waited their turn to tie off on the point exiting the cove and bag the net from there out in a large semicircle until it was time to untie from the beach and cinch up both ends of the bag. Set after set was made, with the catches coming over the stern rollers dominated by large chinook. A dozen or two fish per set looked to be the average. Those numbers were puny compared to the catch per set for the schooling species like sockeye, but large red chinook fetched premium prices that easily justified the effort made to target them. Being an ardent back eddy cut plug moocher at the time, I was depressed, to say the least, by the straining of all those fish out of their migration corridor only days away from where we often sought them in Seymour Narrows.

Not many years later there was a ribbon boundary established eliminating beach tie offs and restricting seine fishing to the middle of the channel through Johnstone Strait. Fishing times were also shortened dramatically. Those measures were intended to focus the commercial harvest on the schooling species, sockeye, pink and chum salmon, and minimize the impact on chinook salmon that were prone to travel along the beaches. Sadly, by then, decades of long seasons and beach tie offs had already taken their toll.

The thing that struck home from observing the seiners in operation all through the areas adjoining the estuary of the Tsitika was the impact on its summer steelhead and many other similar stocks that would have been moving through those waters coincidentally. For most of its history the seine fleet operated from early spring until late fall in Johnstone Strait. With the Tsitika at the epicentre of seine activity in Johnstone Strait, and considering how the fleet went about its business for so many years before anyone ever tried to put Tsitika steelhead on the map, one can only wonder what might have existed before the seines.

A bag of fish coming over the stern roller of Miss Cyanea, the smallest of the seine vessels present at the time, off the tip of Growler Cove on July 11, 1978. Ian Carruthers, an ardent steelheader, salmon fishing guide and friend of the author, was a crew member on this vessel

The Tsitika was not remarkable for its fishing opportunity. In fact it was quite the opposite. For its length it had very little fishing water. It didn't take more than a fly-by to reveal that there was a dearth of fishing water upstream from Catherine Creek. Beyond there were only three or four pieces of water well enough configured to suit steelhead destined to occupy them for long months between late summer low flows and springtime spawning. Those pieces were isolated by long sections of shallow cobble and boulder riffles that were always dreaded by whichever crew members drew that section of river to slither and stumble through during snorkel surveys. The results of those never varied. Almost every summer steelhead in the system was accounted for in not more than a half-dozen runs and pools. The Catherine Creek confluence pool consistently held more fish than any other piece of water on the river. For that reason it was closed to fishing before the logging access got anywhere near it. With that deducted from the total available fishing water there wasn't much incentive to put out the time, energy and/or expense required to fish the Tsitika when it was still in its unmolested state.

Tsitika steelhead will best be remembered, not for their contribution to their home river fishery, but for the fishery they supported in the Campbell River. In the heady days of the federal government's Salmonid Enhancement

Program in the late 1970s and early 1980s, opportunities abounded to implement programs never before possible. Hatcheries proliferated, as anyone who fished east coast Vancouver Island streams like the two Qualicum rivers, the Puntledge, the Campbell/Quinsam and a few others serviced by new and expanded facilities on each of those systems will remember. One of the proposals I advanced was to create a summer steelhead fishery on the Campbell River by using Tsitika River fish as the donor stock. It was not without deeply rooted personal trepidations that I brought this forward. After all, this was Haig-Brown's home river, one with a storied history built around its wild fish. The native Campbell River summer run steelhead were all but extirpated, however, so it didn't seem the worst idea ever to try to create a fishing opportunity in a river with ready access and sufficient summer flow to provide one.

The steps in the process were the most contorted of any fish culture program I have ever been party to. First it was all the disease profile data that had to be examined. That meant collecting a few adults from the Tsitika and submitting them to the pathologists of the day to do their work. A clean bill of health cleared the way for the transplant application and the requisite screening by the federal and provincial governments' committee of experts on all such things. Once secured, the business of collecting brood fish and holding them for the months between collection and maturation followed. That involved some pretty sophisticated equipment and techniques developed specifically to address the nuances of the Tsitika/Campbell program. Spawning and early rearing took place in a carefully chosen groundwater hatchery facility in Duncan, following which the advanced fry were transported all the way back up-Island, well past their river of origin, to net pens in Keogh Lake near Port Hardy. In the spring the smolts were loaded into tank trucks and on the highway once again, this time to the release site in the Campbell.

The first three or four releases of Tsitika smolts to the Campbell produced precisely what had been hoped for. Like their parents, returning fish were mostly of the three-ocean variety, and their numbers reflected survivals that made any hatchery program look attractive. The first smolt release was in May 1982, with the first returns from that group arriving in 1984. Anglers were quick to tune in. By 1985, when the highly anticipated three-ocean animals arrived, Campbell River, a fishing community in a league of its own, was abuzz with tales of summer steelhead pushed by big water stripping fly reels to their arbours. With the Campbell and lower Quinsam rivers still winter steelhead mainstays, the relentless pressure and the fact that hatchery fish were legal quarry meant that few ever survived to spawning. No matter, it was never expected that the imported summer fish would sustain themselves. I liked to think Haig-Brown would not be completely disappointed with a summer steel-head fly fishing opportunity like no other on all of Vancouver Island in a river otherwise on its knees.

Unfortunately the euphoria that developed around the Tsitika summers in the Campbell was little more than a honeymoon. Nothing changed in terms of the program delivery details, but the ocean survival rates that greeted the first few broods did not endure. Five years in, the returns plummeted. Another five and the parent Tsitika stock had diminished such that stealing from it to support a hatchery program elsewhere was raising serious concerns. Fish culturists and fishery managers of the day explored every science-supported approach to sustain both the parent stock and provide sufficient brood fish to keep the popular Tsitika/Campbell program alive, but, in the end, it was abandoned. From the time of that decision to the present the angling closures on too many east coast Vancouver Island streams have discouraged the veteran anglers and eliminated any hope of replacing them with a new generation. Throw in downward-spiralling government budgets and abandonment of anything approaching steelhead advocacy by those still left to call themselves fisheries managers and we can be sure Haig-Brown's home river has seen the last of a summer steelhead fishery.

Some impressions left by the Tsitika before logging, before access and before the rise and fall of a hatchery program that fed off it deserve to be on record. For me the greatest of them was the stability of the river channel and the physical quality of its fish habitat. No valley bisected by roads and altered drainage courses with any significant amount of its virgin timber removed can absorb major coastal rainstorms without effect. The same rain that would tear apart its heavily logged contiguous neighbours, the Nimpkish and the Salmon, would see the Tsitika overtop its banks and wash its riparian zones thoroughly. It could take two or three days following a flood for those other rivers to drop and clear to the point where an angler might have confidence his offerings could be seen by his quarry. On the Tsitika the river might be high enough that getting past the flooded riparian zones to fishable water was a challenge, but I never saw it when water clarity diminished one's expectations of catching fish if you did.

The other notable contrast between the Tsitika in its natural state and all those other once fishy streams bordering it was the rate of rise and fall when the big fall and winter storms descended. It's a pity there aren't historic flow

Images of the brood stock collection equipment involved in the Tsitika–Campbell River summer steelhead enhancement program. October 24, 1980 and November 5, 1982.

records on any of those places to illustrate the point, but the fact remains: the post-logging hydrology sees highly compressed spikes on the hydrographs. The same water that flushed through the Salmon River, for example, would be sponged and held by the Tsitika over a much more protracted period. The window of prime fishing conditions on the unlogged Tsitika would last for several days following the worst flood of a season. Elsewhere that window is getting progressively narrower. Judging from the aerial views of the Tsitika available on Google Earth, I'd wager the same circumstances now prevail there too.

Some observations made following a day of snorkel surveying the pristine habitat of the Tsitika in mid-August 1977 have been tucked away in my mental archives ever since. For 40 years since they have served as a personal benchmark.

We knew from previous experience what to expect in terms of the snorkel survey ahead of us on that August day. The fish would be distributed in predictable fashion and the time required to cover the river was well understood. With all that known before we choppered in that day, our plan was to leave enough time between completion of our swim and the flight back to the Campbell River base to do some exploratory underwater photography. I had recently purchased a Nikonos camera and we had the perfect spot picked to try to capture some images of the object of our affections.

The stage was set. There was a group of ten fish resting quietly in close quarters in a lovely little run bounded on one side by a sloping, moss-covered bedrock shelf and on the other by a clean cobble beach. It was an ideal fishing location as well. The water speed and depth couldn't have been better and the surface was rippled just enough to offer the security that kept the fish from becoming too flighty. I was able to belly my way quietly along and down the bedrock bank to water's edge. With mask and snorkel donned I inched my head and shoulders into the water slightly downstream of the last fish in the bunch not more than ten feet away. Several pictures later the next stage of our experiment unfolded.

One of my staff, Gary Horncastle, was as accomplished an angler as he was a river swimmer. Rod in hand, he took up position on the shore opposite me and perhaps 50 feet upstream. Large rocks submerged in the middle of the run just above the group of ten marked precisely where the fish were. Gary presented a tiny piece of bait looped into a small hook on a light, shot-weighted leader as delicately as possible well upstream from the rocks and ran a drag-free drift of his tiny, natural cork float to the fish behind while I observed through my diver's mask. Cast after cast the pattern was the same. Long before I could see either the float or the bait, the group of fish became visibly agitated.

Several of them would break formation and nervously circle about but never moving more than a few feet outside their original holding perimeter. On each cast one fish would move forward of the group, intercept Gary's bait as it came into view and then drift passively downstream, alternately inhaling and exhaling it until reaching the main group of fish. No taker ever went downstream past the group. Instead the bait was simply allowed to drift on by.

After three or four passes of Gary not being able to detect the slightest movement of his float as the bait repeatedly disappeared and reappeared from the mouth of a fish, we conversed across the river and decided he should strike when I shouted through my snorkel to do so. I recall it took several more casts to get our choreography down, but he finally reacted to my muffled utterances as I watched his bait disappear one more time. He struck, a fish was hooked and I got to observe the struggle, camera ready. I managed to get a couple of shots of the fish doing its thing, but trying to get the focus organized on a moving target in less than perfect light conditions didn't result in anything special. What was more capturable was the behaviour of those fish under angling conditions and the perfect habitat that graced that river at the time. In all my days afield I've never encountered another situation where I could get so close to adult steelhead and have the opportunity to observe their response to an angler's bait. It took less than an hour to learn more than a fishing rod, a library worth of science and the infamous Internet combined could ever teach.

The group of ten large summer steelhead observed and photographed in the Tsitika River on August 16, 1977.

8

SPORT FISHERY "MANAGEMENT" AT ITS WORST

Hearkening back to my early days as a resident of Vancouver Island, I remember well the Harding dinner table conversations about the Stamp River with Ted Sr. and his lifelong angling companions. They were the bridge between the generation of ardent flyfisher General Noel Money and his esteemed angling colleagues, which is credited with discovering the fishery in the 1920s, and my own. There was much talk of the beat from Black Rock through Money's Pool to the Ash River confluence. The focus was on summer steelhead, which accumulated there in numbers that supported catches I'd not imagined possible during my apprenticeship years on the other side of what was once called Georgia Strait.

I never developed a love affair with the Stamp, and even less so with the Somass and the Sproat. My early visits to the waters immortalized by Money et al. were important in terms of developing a perspective on what the system was all about and how it ranked among the many other well-known rivers on Vancouver Island and beyond. The fish were unremarkable in my view but the surroundings in which they could be caught were and still are a reflection of past glory. Lake-headed systems and unlogged riparian habitat do wonders for preserving angling aesthetics. The provincial park at Stamp Falls certainly helps in that regard. What doesn't bear any resemblance to the history of the Stamp system is the sport fishing experience that followed implementation of a hatchery program. Lessons learned in that respect ought not to pass unnoticed. But once again, some background is in order.

Among Vancouver Island watersheds, the Stamp system is one of the larger ones in terms of volume of discharge. The Stamp itself is the river upstream from the system's confluence with the Sproat River; downstream from that point, it is the Somass River. The Department of Fisheries and Oceans (DFO) Robertson Creek Hatchery is located near the outlet of Great Central Lake, the headwaters of the Stamp River. The distance from the top end of the fishing water at Robertson Creek Hatchery to the Stamp/Sproat confluence is about 11 miles. In between, the Ash River joins the Stamp just over two miles below the hatchery and Stamp Falls sits roughly eight miles below. The Somass River is slightly more than two miles long before reaching tidal influence. The Sproat River is a mere 1¼ miles long.

In its original state, the Stamp River proper supported discrete runs of summer and winter steelhead. Most, perhaps all, of the summer fish originated from the Ash River, which had two discrete falls that only summer steelhead surmounted. An unanswerable question at this late date concerns the influence construction of the fishway at Stamp Falls had on the distribution of summer and winter steelhead. There is a case to be made that it gave winter steelhead access to areas that were formerly the exclusive preserve of summer fish.[14] Dr. Fred Withler (RIP), then a research scientist at the Pacific Biological Station in Nanaimo and an ardent steelhead angler, suggested to me around 1976 that growth in the Great Central Lake sockeye population was not necessarily due entirely to fertilizer application, as was commonly held. In his opinion, the access improvements realized by fishway construction had resulted in a natural expansion of the sockeye population independent of the nutrient additions. We speculated that the fishway could also have influenced the reproductive isolation once thought to exist between summer and winter steelhead. This speculation was fuelled by the emergence of a run of bright, fresh-from-salt steelhead at the Stamp Falls fishway in late October and November during the late 1970s. That run prompted me to search out experience elsewhere with the mixing of summer and winter steelhead.

I was fortunate to have my Oregon Department of Fish and Wildlife namesake, Oregon Bob, to assist in recovering a technical paper I vaguely recalled having read around the time I'd been conversing with Dr. Withler. Bob went to work and eventually dug up a copy of what I had remembered from 40 years earlier. The gist of it was that Oregon fisheries biologists working on the Siletz River between 1968 and 1973 had crossed summer and winter steelhead, whose progeny's river-entry timing was perfectly intermediate between that of the parent groups. Once again, this is nothing more than enlightened theorizing about what may have happened over multiple generations of potential mixing of Stamp River summer and winter steelhead, but it is an interesting sidebar nonetheless. Such fish were never acknowledged by the likes of General Money in any historical records, either – although that isn't definitive, given that references to his fishing relate to summer excursions.

One can't appreciate the evolution of the Stamp/Somass steelhead fishery without a basic understanding of the relationship between the federal DFO and the provincial Ministry of Environment (MOE) relative to hatcheries. (The provincial agency has undergone a number of reorganizations and name changes over the years, but for most of the history of Robertson Creek

Hatchery it was MOE.) That relationship began with the implementation of DFO's Salmonid Enhancement Program (SEP) in the mid-1970s. The program's original objective was to double the abundance of Pacific salmon and restore commercial fishing harvests to something akin to what they had been years before. Now, an astute representative from the province, Ron Thomas – who may have been outnumbered but was never outgunned in the federal boardroom discussions during the developmental stage of the SEP – pointed out that doubling salmon abundance to effectively double the salmon catch would have dire consequences for steelhead. Ron drove home the fact that the net fleets up and down the coast had already done serious harm to steelhead stocks and insisted there would have to be some form of compensation. Besides, the province was responsible for the water licences that were required for DFO to operate its hatcheries and that could be used as a trump card. So began the hatchery steelhead era, not just at Robertson Creek but in numerous other facilities on Vancouver Island, the Lower Mainland and as far afield as Kitimat. It was never a marriage made in heaven.

Details around the true beginning of hatchery steelhead production at Robertson Creek are sketchy. The hatchery itself came well after the original fisheries enhancement facilities had been constructed at the site. I often wonder if a single one of the guides or anglers lapping up the hatchery product today knows the facilities now in operation originated as a pink salmon spawning channel that was officially opened on November 4, 1960, by then Minister of Fisheries J. Angus McLean. The hatchery came years later, after the pink salmon experiment had failed. I had reason to believe the typically autonomous hatchery staff had been doing their own little experiments with steelhead before any provincial authority arrived on the scene – that reason being an accumulation of summer steelhead at the hatchery outfall at Boot Lagoon well in advance of any formally sanctioned steelhead production. Coincidental, perhaps, but curious nonetheless. To reiterate, the traditional heartland of the summer steelhead fishery in the system was the Ash confluence area where fish congregated awaiting preferred conditions in the Ash itself. General Noel Money and friends, who were responsible for the earliest formal records of the Stamp/Ash steelhead sport fishery, did not spend their time around Boot Lagoon.

Regardless of any prior meddling, by the early 1980s we provincial people with the steelhead management mandate had designed a smolt production program that met the highest science standards of the day. We had also made a major effort in evaluating every step of the smolt production process to ensure the best possible fish were sent seaward. The early releases were 100 per cent summer steelhead, all of which were derived from wild parents – or should I say parents whose adipose fins were still intact. Collecting those fish required no effort, given their willingness to swim right into the hatchery trap. Every smolt also bore a wire tag coded according to the specifics of rearing group details such as the diet it was fed, size, release date and location etc. No other hatchery steelhead program in British Columbia saw anywhere near as much effort dedicated to optimizing the return on investment. Tragically, tagging became a province-wide fiscal casualty short years later when the cost of the tags plus the cost of their application became unaffordable. So much for fine tuning hatchery production based on measured results.

Summer steelhead production was de-emphasized as the hatchery program evolved. There was relatively little interest in those fish given the abundance of ocean salmon fishing opportunities that proved far more attractive to most of the client base that purchased steelhead fishing licences. Summer steelhead were also subject to interception by commercial and First Nations nets, whereas winter steelhead escaped that fate. The propensity of hatchery summer fish to swim into the facility and remove themselves from the river fishery did nothing to improve their efficacy, either. These were the factors underlying the decision to reallocate most of the hatchery resources to winter steelhead smolt production.

The winter steelhead fishery in the Stamp and Somass Rivers was hardly one to write home about in the years immediately preceding the hatchery program start-up. One can only speculate what role the discharge of pulp mill effluent into the confined waters of the Somass estuary and upper Alberni Inlet may have played in the abundance of and fishing success for preferred species like steelhead, chinook salmon and cutthroat trout in those waters by that time. What is certain, though, is that the anecdotal sport fishing records from the 1950s bear no resemblance to those of the following two or more decades. One of my neighbours during our years as riparian residents on the Englishman River worked in that pulp mill for almost three decades before retiring in the 1980s. He told me that under cover of darkness they frequently discharged materials known to be damaging, if not toxic, to fish and fish food organisms. One only has to make a cursory examination of the estuary and nearby approach waters to the Somass to appreciate that the pulp mill location and the acre upon acre of nearshore areas covered by log booms couldn't possibly have done fish any favours. Sawmilling activity on estuarine uplands predated the pulp mill by half a century or more.

In the early days of SEP, when money was flowing at levels we came to appreciate only years later when it dried up, fellow staffer Gary Horncastle resided in Port Alberni. His primary winter job was to conduct a thorough creel survey on the system. Fishing effort at the time was concentrated on a few runs of the Somass and around the Stamp Falls pool. The remainder of Stamp and the Sproat rivers saw very little activity. The cumulative total effort on the Stamp, Somass and Sproat combined never matched the effort expended on other, more popular streams of the day such as the two Qualicums, the Campbell and Quinsam, the Nanaimo and, especially, the Cowichan and the Gold. The other thing of note was that we equipped Gary with a jet sled to assist him in familiarizing himself with all the access points and to canvass the areas where angling effort was highest. That was the one and only boat of any description on the water at the time.

First returns of winter steelhead arrived without much fanfare in 1983. No concerted effort was made to promote the program, so it took a year or two for the angling community to tune in to what was available. The coincidental catch-and-release measures to protect wild steelhead all over Vancouver Island meant that harvest opportunities gravitated to hatchery streams. The growth in fishing effort on the three-river system was phenomenal by British Columbia standards: by 1986 it was more than five times what it had been in the decade before the first hatchery steelhead arrived. The catch increased similarly as a result of both a relatively large supply of hatchery steelhead and the multiple captures of wild fish. As the fishing traffic increased it didn't go unnoticed that a boat was the ticket to the numbers game. My own boat, first launched on the Somass in 1983, was in the vanguard in that respect. Within three years, a half-dozen jet boats had become frequent flyers on the system. They were accompanied by as many other boats, from lake craft with kickers to car toppers and inflatables with or without power. Problems were emerging. I left the Island for Skeena country in 1986, one fishing season before the arrival of the first guides. What transpired over the next several years provides the consummate lesson in what not to do.

The Coles Notes version of events goes as follows. Emerging awareness of a steady supply of harvestable hatchery winter steelhead meant intensified competition for the best fishing water. Guides – though not the only boaters – were a major component. Riparian property owners became agitated over

noisy jet motors that roared up and down past their heretofore quiet residences with steadily increasing frequency. As the race to be first escalated, the
motors got bigger and the starting times advanced deeper into the pre-dawn.
Sympathetic civic administrators responded and invited new restrictions limiting motor size to 10 HP or less pursuant to the boating regulations of the
Canada Shipping Act. That eliminated noisy jets because those units are not
available for motors under 25 HP. However, the new rule didn't stop one particular guide, who deliberately operated an 80 HP jet motor so that he could
be charged and proceed with challenging the regulation. The charge came on
December 8, 1994. On November 22, 1996, presiding Judge Joe rendered his
decision in provincial court in Nanaimo. In his statement of record the honourable judge held that two factors were grounds for exempting the accused from
the provisions of the horsepower-restriction regulation. First, the province had
issued him a permit (i.e., an angling guide licence) to partake of an activity
that had been his sole source of livelihood for his entire adult life. Second,
the judge accepted that the accused depended for sustenance on steelhead
caught while guiding. For these reasons the judge ordered that the charge be
dismissed. His decision was not appealed.

The transcript of Judge Joe's decision bears testament to how woefully ill
prepared the province was in dealing with the case. Instead of sending a Crown
witness conversant in policy and regulations, with the authority for their application, it sent its lowest-ranking fisheries staff member. The result was that
major features of the circumstances were never raised, much less challenged.
For example, evidence entered and quoted by the judge stated the accused had
"provided river guiding services for about 17 years." Assuming the 1994 date
of the charge was the starting point for the calculation, that meant the accused
began guiding in 1977. He would have been 8 years old at the time, 11 years
younger than the minimum age to qualify for an angling guide licence and six
years before the hatchery steelhead returns he was singularly focused on for
his livelihood and sustenance had even begun. Furthermore, the accused was
not a guide during the period that preceded the charge. He was an assistant
guide for his legal-age sibling who, according to the judge's record of sworn
testimony by the accused, had never acted as a freshwater fishing guide and
had nothing to do with his guiding business other than facilitating cheaper
liability insurance and occasionally booking a customer for him. Additional
records indicate the accused's sibling was first licensed in 1989. For the two
fishing seasons prior to that, the accused held an assistant's licence under his

father, whom no one I know had ever seen on the river during the period of concern. None of this was revealed in court despite the ready availability of such facts. I'll add that the provincial regulations governing angling guides state that the guide for whom the assistant is licensed must be present on or near the water during substantially all times that guiding is occurring. Once again, no one was paying attention.

Judge Joe's decision contained one other interesting revelation. Arguably it is tangential to the greater issue of the motor size restriction and its implications, but it speaks volumes about how guided fishing was and still is prosecuted on the Stamp/Somass system. The record of testimony noted the accused stated that, on average, he was responsible for 40 per cent of the fish caught by the occupants of his boat. In other words, the guide fishes, hooks the fish and then hands the rod to the client to bring to net. According to the sport fishing regulations, the person who hooks a fish owns it, not the person who may ultimately take the rod and reel it in. The regulations further state that once a legal limit has been landed on any given day, the angler must cease fishing for the remainder of that day. In the topic case the accused would have exceeded the daily limit more often than not. One can only wonder why he was exempt from that regulation too.

Independent of the failure of the court to recognize any of the above, the critical point in Judge Joe's rendering was that he didn't exempt all comers, only the single guide who had contested the regulation. The BC Wildlife Federation seems to have been the only organization concerned that the decision could potentially apply to guides other than the one who appeared in court. A quote taken from the response by federal Minister of Fisheries and Oceans David Anderson to a letter written by then BCWF president John Holdstock (RIP) is the only evidence of any attempt to confirm that Judge Joe's decision would not throw the door wide open:

> In our opinion, that decision only grants an exemption to the specific individual concerned, and for very specific reasons, and does not suggest that all fishing guides are now exempt from any or all restrictions from here on. Consequently, enforcement of any restriction should, in our opinion, continue in the same manner as it was being carried out before that court decision, leaving it to the court to rule on special cases which may be brought before them, from time to time.

I am unable to find any record that the province ever moved in accordance with Minister Anderson's advice. In fact, by the time of the Judge Joe decision, the Stamp/Somass system was populated by "about 20 guides," all of whom had convinced themselves they could behave just as their compatriot had unless or until charged.[15]

In early October 2013, I revisited the upper Stamp River to refresh myself on the angling experience available on the waters immortalized by General Noel Money and friends. Flagrantly violating the law in that I didn't have the guide licence that would legitimize my modest little jet boat, I launched in Boot Lagoon at the hatchery site and proceeded down the Stamp River past the Ash confluence and on to Money's Pool and Black Rock. I wasn't particularly interested in fishing but I did bring along a Spey rod to enjoy some casting around Money's while I waited and watched. Over the course of the day, I encountered four guides in boats considerably larger than my own and sporting double the horsepower. Floating tackle shops would not be an unfair description of them, judging from the number of rods rigged and ready. Each boat had two clients aboard. None of the guides ever shut off their oversized jet motors while in my field of view; instead, they just idled back and forth across every piece of potential holding water, often fishing right on top of fish they could plainly see. If they didn't "hook up" on a first pass they just powered up and did it again with a different set-up or moved on to the next spot. Apart from the social aspects of the circumstances that now prevail, has anyone even contemplated what such practices are doing to the fish? So much for the storied haunts of General Money.

In the final analysis, the Stamp/Somass stands as the benchmark for all that can be done wrong in managing a steelhead sport fishery. Here the taxpayer had financed a hatchery program successful enough to turn the receiving waters into the most heavily fished system with the highest steelhead catch on Vancouver Island (and one of the top three in the province). Short years later, all that the sport fishery was intended to be was effectively handed over to guides on a silver platter. How else could a reasonable person view it when a select group of what are essentially commercial fishermen enjoy such disproportionate benefit from a publicly funded supply of hatchery fish (as well as all the cohabiting wild fish)?[16] What other conclusion could one reach when the province took no visible steps to address the obvious inequities that arose from Judge Joe's decision, even after the federal government's interpretation of that decision recommended otherwise? Why has there never been an attempt to apply existing regulatory tools, such as no fishing from a boat or no power boats, much less to develop new regulations to level the playing field?[17]

The institutional negligence that gave rise to present circumstances is unpalatable, to say the least, but those who expect that the public bounty will persist may be in for disappointment. This statement requires a bit of discussion that makes for less than exciting reading, but the facts are certain to be lost and forgotten otherwise. Bear with me.

When the hatchery program for winter steelhead began, the strategy was to capture newly arrived fish from the Somass River. The collection period targeted the earliest fish possible because their timing ensured the longest pre-spawning freshwater residence and therefore the longest potential contribution to the sport fishery. Smolts were released below the Sproat/Stamp confluence to keep the hatchery fish as separate as possible from their wild counterparts, which typically moved upstream into the Stamp far beyond the Sproat confluence. The plan was laudable but the ratio of hatchery to wild fish plus the relentless competition from guided anglers in boats made it ever more difficult to obtain the desired wild steelhead brood stock. Compromises began to be made.

Agency staff opted to use volunteers to assist. Guides became ever more prominent in that regard. The area of collection kept creeping upstream, thus increasing the uncertainty about river-entry timing of the brood fish. Smolts that had originally been released well downstream began to be released as high up as Stamp Falls because, at the time, it was one area that had historically provided access and angling opportunity but wasn't being fished by jet boating guides. It seemed logical to try to cater to someone other than guides, so adults homing to a smolt-release site upstream from the guide gauntlet was entirely reasonable. The desired objective – producing more fishing at Stamp Falls – was achieved, but the guides were quick to move in and dominate that scene as well. As if monopolizing every decent piece of water in their original turf downstream wasn't enough, now they were racing to the falls pool at dawn to idle their 100 HP motors up and down, back and forth, easily outcompeting anyone who made the effort to hike into the site. That made it even harder to obtain the requisite brood stock.

The next step in the downward spiral came with collecting brood stock from the top end of the Stamp system, immediately downstream from the hatchery. At that point any certainty around the river-entry timing of the brood stock was gone. The time sump that brood stock collection had become for agency staff was alleviated somewhat by closing a section of the upper river to all angling during the winter steelhead season, which substantially improved the catch-per-unit effort for those certified to partake. Convenience and short-term cost effectiveness prevailed while the longer-term biological risks inherent in homogenizing the system's steelhead were allowed to fade from view.

Several other factors have come to bear on the hatchery program of the present. The business of a shared responsibility for producing smolts at the Robertson Creek facility is not what it was. Provincial fish in federal hatcheries has become a sticking point at all federal hatcheries. I reiterate: it was never a marriage made in heaven. Steelhead smolts of a desired size and quality are far more taxing to produce than chinook or coho. No one in a federal facility will ever admit to it, but I learned from career-long experience that steelhead rarely receive the same priority as salmon. How many times did I hear complaints about the cost of food, the time involved in sorting and grading to meet size and time of release guidelines or the sheer physical space and attendant maintenance recommended by provincial people better versed in steelhead biology and culture than the federal people charged with growing the fish? When budget cuts just keep on coming, morale among hatchery workers nosedives. Extras like steelhead suffer. The records speak for themselves in that respect.

Uncertain origin of brood stock, mixing of hatchery and wild steelhead in natural spawning areas, decline in smolt quality, stocking of surplus hatchery fry into what was once the best wild steelhead habitat in the watershed, sacrificing of reproductive isolation of summer and winter steelhead by manipulating migration obstacles, compromise in wild steelhead smolt production because too many hatchery reared juveniles that fail to emigrate and outcompete them for space and food, chronic disease issues, persistent low ocean survival, and predators who have adapted to a hatchery-created twelve-month smorgasbord – all these things, singly or in various combinations, forebode a very different fishery. That scenario is not exclusive to steelhead.

The other force that has major implications for steelhead is the First Nations fishery. In earlier times, fisheries enforcement staff had tools and support for addressing gillnets set in rivers. I have personal experience in removing nets and I contributed to legal action taken against those who set them. The First Nations community in the Port Alberni area were well aware of the presence of enhanced steelhead and increased their harvest effort accordingly. The measure of their understanding of the legality of it all was the fact that they only set and retrieved nets under cover of darkness and fished those nets such that the cork lines were sunk and invisible to the casual passerby.

Contrast that with the circumstances that prevail today. Granted, the focus is on enhanced sockeye, but their return timing matches that of summer steelhead closely. The effort now directed at harvesting surplus sockeye has reached absurd levels. Successive waves of gillnets specifically constructed to stretch from bank to bank and from surface to bottom being guided down the river by custom-designed outboard-powered boats on each end of the net is not something the remaining summer steelhead stock is capable of sustaining. No one in a position of authority is making any credible effort to even monitor the impacts on steelhead, let alone do something about it.

On a lighter note, I'll finish speaking of my association with the Somass and Stamp rivers with a fishing story. Dateline: spring 1975. Doug Morrison, the officemate I spoke of in my chapter on Gold River, joined me on an exploratory trip in search of summer steelhead in the Nahmint River at the end of May. We launched my jet-equipped 14 ft. Zodiac Grand Raid at the marina at the lower end of the Somass River in Port Alberni and ventured off down the inlet to our destination. The fishing didn't prove out so we returned early. On approach to the launch site we conferred and decided we should run up the Somass and see what opportunity that might afford.

Now, I had heard of Papermill Dam but I'd never seen it. About 2½ miles upstream from our starting point, we were going through tidal and placid non-tidal water until, all of a sudden, we saw the whitewater cascade over what had to be the dam. It was a pretty impressive sight with the river full from seasonal meltwater. I throttled back and held position a respectable distance below the turbulence at the base of the whitewater while contemplating my next move. Careful observation of the channel revealed a bit of a slot on river left that was slightly less violent and potentially navigable. We decided it was doable if I took a run at it.

Doug sat facing me on a combination gear-storage box and gas tank compartment I had carefully crafted to be wedged tightly between my Zodiac's pontoons. He had a firm grip on the rope strung through the tie-downs along the top of the portside pontoon. At a distance below the wall of water of about 30 yards, I cranked the tiller throttle wide open and made for the narrow chute we had identified. We hit it at the precise moment the standing wave we were planing over collapsed, with the effect of dropping the bow of the inflatable into the vacuum between two standing waves. As the first of these collapsed, and the bow dropped, it was met by the second one coming up directly under us. The boat literally folded right under the seat and catapulted Doug over my left shoulder into the river behind me. He disappeared completely from view for what seemed an eternity but was probably not more than a second or two. Suddenly, an arm thrust itself through the froth – still clutching a rod bearing a beloved Hardy. Years later I referred to it as "Excalibur revisited," but at the time all I could think about was grabbing that arm and wrestling its owner back into the boat. No one would ever accuse us of being at the high end of the IQ spectrum that day, especially after we emptied Doug's waders and lined up for a second attempt at the falls. We made it without incident that time and ran another five miles or so to a piece known as the Girl Guide Run. An afternoon of blue chipping our way downstream from there to the Sproat River confluence produced nary a fish. The return trip over Papermill was uneventful. We never saw another soul on the Nahmint, the Somass or the Stamp that day, nor did we encounter another vessel doing anything related to fishing along our route through Alberni Inlet to and from the Nahmint.

There is one final quote that I believe captures the essence of the Stamp/ Somass steelhead fishery today. I've borrowed it from an address given by Roderick Haig-Brown at the annual meeting of the Steelhead Society of British Columbia on February 11, 1973: "In the end, of course, it is the fisherman who will settle the matter. He can choose to catch a fish or two shoulder to shoulder with his brothers in a pool immediately below the hatchery intake, and imagine himself a steelheader. It is a pleasant conceit that may yield satisfaction for a brief span. Or he can decide he really wants to go steelhead fishing."

The author running Papermill Dam on the lower Somass River on November 27, 1983.

One of the very early hatchery winter steelhead caught in the Somass River by the author's friend Rob Hobby on November 27, 1983.

More examples of hatchery steelhead caught by companions of the author in winter 1984 and 1985.

Examples of the boats and nets now commonly used by First Nations fishers intent on harvesting enhanced sockeye in the lower Somass River, June 23, 2014.

The hidden gillnet containing five winter steelhead retrieved from the lower Somass River by the author on November 27, 1983.

9

A DIFFERENT WORLD

John Fennelly's *Steelhead Paradise* was my Christmas gift from an uncle in 1963, the year it was published. At the time, Fennelly's description of the Sustut and Johanson country had me spellbound and dreaming of a day when I might experience such a world. The seed, once planted, took a few more years to grow, but by the mid-1970s, I was ready and equipped to head off to the steelhead's promised land.

My introduction came in 1974. It followed from the second trip I had made to the Dean in August of that year. There I met Jimmy Wright (RIP), arguably the most colourful and talented person I've ever encountered in the pursuit of steelhead. He was camped on the river a run or two upstream of where Ian Carruthers (RIP) and I had set up for the back half of our week on the water. Wright's partner on the Dean was Bus Bergman. Over those few days of sharing water and philosophies of all things steelhead, a bit of a relationship developed. I learned that Bergman had spent a fall season or two on the Babine guiding for Babine River Resort builder and owner Bob Wickwire. Bergman's tales of the Babine contrasted sharply with those of Fennelly and were more than enough for me to add that river to my wish list. Fennelly's efforts had been focused on the wrong parts of the river at the wrong times so it was no surprise he never ranked it highly.

Years later, after I had moved to Smithers, I was able to piece together much more of the history of the Babine steelhead fishery. Fennelly wasn't the only one who failed to figure out where to fish and when. The earliest fishermen – who came well before him, beginning in the mid-1950s – had all operated out of the original Norlakes Lodge near Fort Babine on Babine Lake. They boated to the fisheries weir some 15 miles distant and fished on foot in that general vicinity, an approach that always produced occasional steelhead but never the motherlode. Rainbow trout, Dolly Varden and coho dominated the catch in those days. There were some intriguing accounts of catches of those species that illustrate plainly what will never come again. It wasn't until anglers began to venture farther downstream that the Babine's steelhead bounty was discovered. The pivotal moment was on November 2, 1955, when

Tom Stewart and his two business partners, Mac Anderson and Ejnar Madsen (RIP), hired two local First Nations men to pack their gear in a prop-driven, flat-bottom Heritage Classic lake boat down the Babine to the Nilkitkwa River confluence.[18] There they discovered steelhead in numbers that gave rise to the first steelhead lodge, Norlakes, built one bend below the Nilkitkwa on the opposite side in the mid-1960s. That was the genesis of the modern-day steelhead fishery.

The Babine was not on my agenda in 1974. It was clear from the Dean encounter with Bergman that the good steelhead water was inaccessible in the absence of a jet boat, and I didn't have one then. What I did manage that year was a trip to the Kispiox to join Jimmy. His home at the time was The Creamery in Telkwa, long before it became the centre of a guiding operation with a checkered history. My friend Mike Whately, a former supervisor and another mentor from my early days in the fisheries business, was in charge of the fisheries program for the Skeena country. He had hired Jimmy and set him up at Olga Walker's steelhead camp right there in the centre of the steelhead universe. I spent a few days on-site living out of my pickup and canopy, parked beside the cabin Olga had reserved for Jimmy. Those few days were never directly related to the Babine but they had a major influence on all that followed. Thus, the following digression is required, and well deserved.

My timing in 1974 was all bad. It was mid-October and the rains had descended. The Kispiox was unfishable, as was every other river save the Bulkley upstream from Telkwa – and even it was marginal. No fishing left time for talk, though, and there was an abundance of that. What sticks with me so clearly all these years later is the passion Jimmy Wright had for steelhead. He was far ahead of his time in terms of preaching the gospel of catch-and-release. In fact, he was the subject of numerous complaints to his supervisor in Smithers over his aggressive sales pitch to the Kispiox devotees who were still all about wall hangings. Imagine the quintessential religious zealot fly fisherman in the face of an old-time trophy hunter telling him he was unwelcome on the Kispiox if he intended to kill steelhead. My favourite memories of Jimmy that year were his stories of repeated encounters with Bob York (RIP). Anyone familiar with the Kispiox through the final three decades of the last century will know something of "Steelhead Bob." If not, search out a copy of Ben Taylor's book *I Know Bill Schaadt*. Bob York and Skeena steelhead is as close a parallel to Bill Schaadt and northern California chinook as river fishing in BC is ever going to see.

I didn't meet Steelhead Bob in 1974. In fact, it was well into the 1980s before I did. By then he was known far and wide as the ultimate steelhead fly fisherman. The transformation of the man described by Jimmy Wright in 1974 was remarkable. Jimmy had given me one of York's business cards that advertised his services as a drift boat guide on the Olympic Peninsula streams. Plugs, not fur and feathers, was the name of the game. York was every bit the equal of Wright in terms of passion and aggressiveness, but he hadn't yet succumbed to the religion of fly fishing. I can only imagine the theatre that must have prevailed when Jimmy and Bob encountered each other on the Kispiox that year. Jimmy avoided being drowned or fired but it was his last job with the Fish and Wildlife people. York's conversion went unnoticed by all but a handful of Kispiox veterans. The new age crowd held to their belief that he was born with a fly rod in his hand and that he would sooner kill a gear fisherman than a steelhead. I was thankful for the Kispiox experience in 1974. It gave me some valuable background I never would have acquired otherwise.

One year later I had the boat that was going to get me to the Babine water described by Bus Bergman. With a year of Vancouver Island river experience under my belt I headed to the Babine in early October 1976. Fennelly described driving into the lower Babine in 1955 on the road constructed a year earlier to access the Babine slide. He referred to it as the most excruciatingly bad piece of road he had ever attempted to travel in an ordinary automobile. Access to the upper Babine was likely not quite as bad, but in several spots it was less risky to forge ahead than to attempt to turn around. A wet fall in Skeena country that year had made logging roads, especially newer ones not yet gravelled and packed down by heavily loaded trucks, a nightmare. It was touch and go for an old half-ton Ford to slither through that mess, but four hours after leaving Highway 16 I arrived at the fisheries weir. The same road today can be driven in about a third of that time.

The only road access was by the logging bridge near the river's outlet from Nilkitkwa Lake and 300 yards downstream from the weir. That crossing and the road beyond were all about salvage logging after a major forest fire in the Nilkitkwa drainage, in areas well out of sight of the Babine River. No one was using what remained of the 20-year-old Babine slide road that paralleled the canyon-encased lower river, so even the short reach of decent-looking steelhead water near the Skeena confluence remained virgin. Jet units for outboards had arrived, but the only ones in operation belonged to the two guides. Anyone could get a jet boat to the weir, just as I did, but there was enough tricky water between there and the Nilkitkwa River confluence two miles downstream to discourage those who dared test their water-reading skills. The guides weren't above trying to dissuade anyone else they saw with a boat at the launch site near the weir. In part, that was to avoid having to rescue stranded anglers, but maintaining their exclusivity was also part of the equation. Drift boats were off limits – there was no takeout until the Skeena 60 miles distant, and life-threatening canyon water well upstream from that. Inflatables capable of handling that water hadn't yet arrived. The other card nature played on the Babine was grizzly bears. An abundance of salmon kegged up in the reaches downstream from the fisheries weir was a major draw for bears. The heavy presence of grizzlies all around the river's only point of public access was a major deterrent for anyone intending to walk the old trapper's trail downstream on the river's right bank or camp in the vicinity. Elsewhere the impenetrable streamside bush made hiking along the riverbank a nightmare.

At first light on the day following arrival, I assembled my Zodiac, mounted the 40 HP Mercury jet and loaded up gear and provisions for some serious exploration. The last opportunity for communication was 80 miles behind me in Smithers. I have always been cautious of travelling downstream when I know there is no way back if untoward circumstances were to arise. By that time, though, I was somewhat of a veteran of those sorts of adventures, so I never considered that a well-appointed boat reasonably equipped with safety and survival gear wouldn't allow me to deal successfully with whatever may come.

Now, the Babine does not give one a feeling of comfort or inspire confidence. The water leaving Babine Lake is clear but tannin-stained and dark. The lake creates welcome warmth but the air above is cooler than the water at that time of year, which generates fog and mist that lingers through the early hours of the day, if not longer. It doesn't help that the sun is low in the October sky and the river is entrenched and shrouded by spruce forest seldom broken by the splashes of golden aspen that add so much to the scenic backdrops of every other Skeena tributary.[19] All things considered, the experience was less than inviting that first time around. But into the unknown I went, venturing a little farther downstream each day.

By the third or fourth day of my fishing my way downstream, Ejnar Madsen – the original operator of Babine Norlakes Steelhead Lodge – had become curious about the interloper he encountered while taxiing clients back and forth to various fishing spots. No one besides his competition, Bob Wickwire, ever visited those waters in a jet boat, so I was worth some investigation. Ejnar put his boat ashore just above mine and strolled down to meet me. We exchanged greetings and names and sat down for an exploratory chat. While I enjoyed a welcome cup of coffee from his ample Thermos he proceeded to tell me how fortunate I was to be visiting the river and experiencing its navigability when the water was so high.[20] He pointed out that one of the pathways he had observed me taking was a death trap for a jet unit at an average fishing flow. His advice as to the collision-free route through that spot undoubtedly saved me a shoe, if not that year, then certainly in years that followed. I had many other encounters with Ejnar Madsen in later seasons, all of them as friendly as that first one. He was a good man and a credit to the river. I never quite understood why he was so accommodating to one who represented little more than competition, but I like to think our original streamside chat satisfied him that I was a friend of the fish that were his livelihood.

Long-range September view of distant mountains from the run known as "Log Jam" on the upper Babine River.

There is one postscript to my original Babine adventure: the news con-
veyed by my mother when I checked in with her after my return to Smithers
and access to a phone. Roderick Haig-Brown had passed away on October 9.
The long drive south that followed was filled with reflections on his wisdom
and our discussions the year before.

The fishing environment along the Babine when I first visited it in 1976
was unlike anything I had ever experienced, but the overriding impression
I took from that trip was how disappointing the performance of its fish was.
Here we had a lake-headed river with water temperatures as good as they get
for steelhead fishing, a relatively high gradient river with sufficient volume
to carry any of its many large fish out of runs where they were hooked down
through water where they could never be followed. Surprisingly, that never
happened. It took me a few years to figure out that the biggest issue in that
respect was that the fish were essentially at the end of the line in terms of
upstream migration. They were sitting ducks for guided anglers, who applied
relentless pressure throughout the season. I should add that fly fishermen then
were as scarce as hardware fishermen are now. Multiple recaptures were com-
monplace, and fish enduring that were not the stuff of epic battles. Subsequent
years saw my fishing move ahead, from October to September and even up
into late August in some years. Earlier and downstream proved much better
than later and upstream. Bob Wickwire sold Babine River Steelhead Lodge
and built Silver Hilton and its satellite Triple Header where he did for good
reason. (Wickwire was also keenly aware of the potential for destruction of the
upper river fishery if the logging proposals on the table at the time couldn't be
stopped.) The fact that the Babine grew substantially in volume by the time
it reached the Silver Hilton beats, and that most of the fish arriving there had
never felt a hook, would also factor into fish performance.

It takes many years to appreciate the nuances of any river and fish stock
but I'm satisfied that I knit things together fairly completely for the Babine. The
annual trips I made to the upper end between 1976 and 1986, the numerous
trips made in the 13 seasons thereafter when I lived in Smithers and another
dozen in the years since strike me as an adequate sample on which to base
some conclusions. That initial impression of the lethargic nature of such a high
proportion of the fish caught anywhere near the peak of the season or later has
never changed. I can't speak with such conviction for the lower Babine River
because I fished it on only two occasions. The pattern in which summer steelhead near the limit of their upstream migration are outperformed by newer

Triple Header Lodge, the satellite camp of Silver Hilton Steelhead Lodge on the lower
Babine River, September 2007.

arrivals lower down in a river system shouldn't be difficult to comprehend. In
my experience, this pattern is more pronounced on the Babine than anywhere
else I've found.

I should say more about this business of capture rates on what amounts to
a resident trout population. There is no safe haven for steelhead or any other
species present in the upper river. No water is off limits for fishing or inaccessible to anglers during the steelhead fishing season. The reaches below Babine
Lake that gave birth to most of the history of the steelhead fishery are decep-
tively small. The Babine's volume increases significantly with the entry of the
Nilkitkwa River two miles below the fisheries weir, but unless the latter is in
flood, it only serves to add a delectable bit of colour that makes the Babine
more fishable than it would otherwise be. Perhaps a better perspective on the
size of the upper Babine can be had by reference to some other well-known
steelhead waters. Consider the Bulkley near Smithers. Under similar weather
conditions, when neither the Babine nor the Bulkley are in flood or drought, the
latter carries roughly four times the volume of the former. When both the upper
Babine and Kispiox are at optimal fishing flows, their volumes are comparable.
For those more familiar with Vancouver Island, the upper Babine in September
and the upper Cowichan in March (and sometimes April) are very similar in

size. Imagine the Kispiox or the upper Cowichan with eight guide-operated jet boats and another four or five unguided craft jockeying for position on any given day. I have more to say about that in another chapter.

The point to emphasize here is that the Babine steelhead eventually become sedentary creatures, either because they are near their spawning destination or because declining water temperature slows their metabolism to the point where migration ceases. None of the tributaries draws any significant number away from the mother river before the following spring and even then only a handful spawn elsewhere than in the upper Babine itself or a very few small, clear water feeder streams. No fish go beyond the outlet of Nilkitkwa Lake. All of this has been confirmed through radio telemetry investigations. By late October the steelhead population has reached its annual peak. Those present hunker down in well-known overwintering runs, remaining essentially dormant until April and May. Ergo the resident fish analogy. The earlier a steelhead arrives, and the later winter descends, the greater the likelihood of multiple captures. I'm of the opinion that veteran anglers and their guides who do not recognize the vulnerability of these fish are nothing short of negligent. The fact that guides have taken to aggressively marketing fishing into November (the "secret season") because there are no limits on the number of clients they can muster at that time and no fees for either the rod days they account for or quality waters licences for their clients does steelhead no favours. How much of a trophy or memory is a lethargic November steelhead dragged from its overwintering habitat? Fattening the averages and profit margin by preying on fish whose performance has already been diminished by repeat capture and seasonal low water temperatures may be good for the bottom line but it is the antithesis of the quality angling the Babine is known and marketed for.

The changes wrought by Mother Nature on the streams described in previous chapters have not occurred on the Babine to anywhere near the same extent, at least not yet. I have a copy of the original map of the upper river with Ejnar Madsen's handwritten names for productive runs acknowledging so many of his early Norlakes customers. Some of those runs no longer exist because of collapsing riverbanks redirecting flow or log jams being moved by spring freshets, but most of the runs named on that late-1960s map remain intact today. A 100-mile-long natural lake at the headwaters of a system affords a lot of stability. Whereas the natural environment remains largely intact, the same can't be said of the fishing environment.

Good fishing is never kept secret. Even before the Internet and the explo-
sion of information it has provided, the traffic level on the water was increasing steadily. Earlier years saw mostly inflatable craft making the drift from
the road access point at the weir to various takeout points downstream. It
was more common for drifters to be helicoptered out at one time, but more
recently, most ride out the lower Babine canyon waters and take out either at
the Babine/Skeena confluence or at Hazelton. During the late 1970s and 1980s,
I encountered only one other jet boat that wasn't being operated by a guide. I
can't say I was the only person venturing into the hallowed guiding territory
but the competing traffic was very obviously light – a sharp contrast with the
last decade, which has seen a steady increase in the number of jet boats. Gone
are the days of skillful navigation through the bedrock-laden shallows that
Ejnar Madsen warned me about. Now we have the oversized inboard machines
with ultra high molecular weight plastic lining the bottom. Their operators pay
no heed to navigation channels. Straight-line, high-speed travel, bouncing off
boulders and bedrock, is far easier. A screaming 150 to 200 HP inboard jet that
can be heard for a mile, before being seen, does little for one's expectations of
a quality fishing experience on what is heralded as a wilderness river.

A rare encounter with another jet boat operated by a non-guide on the upper Babine
River, September, late 1970s.

The trends evident on the Babine are somewhat depressing when viewed against some of my benchmark memories and experiences. Such is the world we live in. It doesn't diminish the highlights of yesteryear, though. A few favourites are worth sharing.

Grizzly bears are a constant presence along the Babine in most years. One would think I would have had many encounters over more than three decades of camping at various points on the 12 miles of river downstream from the fisheries weir. I didn't. In all the years I experienced only two incidents where there was even the slightest cause for concern. Both occurred at the same location but several years apart. The first was a midday visit while my son, 11 years old at the time, and I were setting up our tent. The bush just beyond clear view crackled with the sound of something approaching. Moments later a grizzly ambled into view; it stopped about 50 feet away and immediately reared up on its hind legs. The three of us stood there wondering what was next for what seemed like a long time but was really a few seconds before I reached for my chainsaw and fired it up. That sent the bear, hind feet up around its ears, crashing a getaway as fast as it could go. We never heard from it again.

The other incident involved long-time friend and fly fishing aficionado Steve Pettit. We were into the second or third night of a trip and nicely tucked into our sleeping bags when a visitor could be heard grunting and rustling around a few feet away from our tent. First response – make noise. We screamed obscenities at the top or our lungs and heard the bruin depart noisily into the bush. Not long afterward, just as we were finally falling asleep, the nervous silence was broken once again. This time the bear was pawing at a cooler containing our carefully stored food supplies, close enough that I could hear its every breath. Loud obscenities worked once more, but this time we reinforced them with a couple of circuits around the tent as I revved my chainsaw repeatedly. Apparently bears don't like chainsaws. We never heard from that one again either.

There is a fish story from that 1994 trip when my son and I were visited by the grizzly. It was early September and only one of the guide camps on the upper river had started up. That left us to fish as leisurely as we wanted in all the water around our campsite. One of the things we always enjoyed about the early season in that part of the river was the presence of a few rainbow trout that keyed in on spawning pinks in a side channel immediately in front of our tent. It was our standard ritual to pick out a decent specimen for dinner on our first night in camp. In the process of searching for one I decided to tie on a #10 deer hair pattern and wade far enough up the spine of the large run just above camp that I could fish the water leading down to the beginning of the side channel. As I pushed my way upstream mid-river I realized there was productive-looking water on either side of the spine. Alternating casts between river right and left, I slowly fished my way downstream. Several casts later the glass-smooth water around my waking fly was sucked into a huge vortex, not once, not twice but three times on that same pass. Frozen in anticipation, I watched as the fish stuck that tiny little fly in its hinge on the third pass. As with most Babine fish, the fight was unremarkable, but I will never forget the image of that 38 in. male attacking that morsel of deer hair. The other memory of that weekend trip was the name we gave the riffly run immediately downstream from my encounter with the deer hunter. Between us (though it was mostly my son) we hooked seven fish on dry flies from that one unlikely looking spot during our weekend stay. Forever after we knew it as Seven Up!

A large dry fly-caught steelhead from an early season trip to the Babine River in early September 1991.

Then there was the buck bug. That story began with Tom Morgan (RIP) (Tom passed away on June 12, 2017), at the time the owner of the Winston Rod Company. Steve Pettit was a devout Winston guy and suggested to Tom he should look me up on his annual trip to fish the Morice River. The following year he did and we spent a wonderful day together touring the Bulkley down

Trout Creek way in my sled. Two things remain with me from that day. One was a line Tom came up with over a streamside lunch. He was comparing the behaviour of Montana trout and juvenile steelhead and the characteristics of the water they chose to occupy. This led to a conversation about the similarities between steelhead parr and adults. He paused for a moment and then said, "You know, steelhead are just little fish in big bodies." How perfect is that? The second is Tom's advice that I search out a W.W. Doak catalogue from the centre of the Miramichi's storied Atlantic salmon country in Doaktown, New Brunswick. The folks at Doak were Tom's supplier of bombers, a few of which he generously left with me at day's end.

As I pored over the catalogue shortly afterward, one particular fly pattern caught my attention. It wasn't so much the fly itself as the magnetism of the words accompanying it: "these are unquestionably the best selling line in our showcases." The black body, green butt edition looked to be the perfect analogue to the popular west coast steelhead fly, the green butt skunk. That was enough to prompt me to order a few, to add a bit more flavour to boxes and wallets filled with conventional steelhead patterns. The test came in late September 1990. "Oregon Bob" and I were preparing to embark on a Babine drift trip, but not before a day or two on the Bulkley to get ourselves into full predator mode.

We launched at the Suskwa Forest Service Road crossing on the lower Bulkley and started downstream. As we got to the bridge I suggested to Bob that the tail of the run right under the bridge was at a perfect height and shouldn't be ignored. Sure enough, I rose a fish to a bomber almost immediately. Several more casts produced no further response. It was time to try the buck bug. One cast was all it took. And so it began.

Two days later we were on our way down the Babine. We didn't do much more than travel and pitch camp until we were about 13 miles downriver from the launch at the weir. All the water to that point was very familiar to us, from numerous previous trips. The second morning of our drift had us at Beaver Flats, as prime a piece of dry fly water as one could ask for. I left it to Bob while I hiked to the run above to fish a tailout that had been very generous to me over the years. The first Babine steelhead ever victimized by a buck bug came to hand shortly after. A picture of that fish graced a BC Federation of Fly Fishers membership solicitation poster for years afterward.

Now, the buck bug story doesn't end there. The next day of our drift found us in "middle earth," the name accorded the five miles of river that is untouched by jet boats coming from either of the two guide camps upstream or from the Silver Hilton/Triple Header crews below. I managed four more fish to hand on dry line and bug before we made camp that evening. The favourable impression was growing. Early the following morning Bob and I hiked back upstream a short distance from our camp to fish an attractive piece of water that we had bypassed the evening before in our haste to get set up for the night. Being able to get to the far side of the river, where we really wanted to be, wasn't in the cards, but we did a bit of damage from the wrong side nonetheless. The sports from below arrived soon after and took up residence opposite us. Later in the day, as Bob and I worked our way downriver, one of the Silver Hilton sleds approached and waved us ashore. At the helm was Bob Wickwire. He recognized me immediately and invited us to bunk in at Silver Hilton for a night or two. There had been a late cancellation among the crew that was due in that week, so there was room at the inn. No one in their right mind would turn that down.

The first steelhead taken by the soon-to-become-popular "buck bug" fly, September 26, 1990.

The first steelhead caught by the author in the Babine River on the buck bug fly
introduced to that water on September 29, 1990.

Babine Bob expressed concern over our plan to drift to Kispiox, citing
"grizzly drop" as a serious hazard at existing water levels. Oregon Bob had
many years and countless miles of major whitewater under his belt, but Babine
Bob convinced us that a better plan was to hang out at his camp for a couple of
days and have his son Jud shuttle us in his helicopter to a truck they needed
driven back to Smithers. In retrospect, I think the bigger reason for the invita-
tion was to minimize our presence on all the good runs that we would be sure
to fish on our journey down through the water – the runs his crews from Silver
Hilton and Triple Header always expected to have to themselves. If he had us
at Silver Hilton, the only way we could go back up and fish any of the good stuff
was via the lodge jet sleds. The deal was that his guides would deploy us after
all the clients were spotted on the best runs or after they had already fished
them. Oregon Bob and I were quite content to comply, given the circumstances.

The dinner gathering one day in was an experience. Naturally the day's events were the primary topic. My diary notes that the six American clients of Silver Hilton, all of them seasoned veterans of the Babine, hooked 15 fish that day and landed nine. Bob and I went four for six and seven for seven, respectively. My seven were all on the buck bug and dry line. So, as two guys who had never fished that piece of river before, and never had the advantage of being first on the water or being given any special attention or advice from our taxi drivers, we still managed to put our hands on more fish than all the paying customers combined. The exclamation mark was the four fish I teased out of the gorgeous piece of water right in front of the Triple Header lodge, where I had asked Jud to deposit me after the four clients housed there had finished with it and moved to greener pastures. Not wanting to offend anyone with our results, Oregon Bob and I kept a low profile. Others talked, we listened. I knew Babine Bob was paying attention to my silly little buck bug, though. When we departed Silver Hilton the following afternoon I left him a couple of them to experiment with and followed up with a few new ones from Doaktown after returning to Smithers. That's how the buck bug was introduced to the Babine. In spite of several items I've come across in recent years in which the source lays claim to either originating or introducing that fly, or takeoffs from it, I'm certain that west coast steelhead had never seen that east coast Atlantic salmon pattern before September 1990. Jerry Doak confirms he has no recollection of ever selling a buck bug to anyone else from the steelhead turf until years after my original order. Thank you for many happy days since then, Tom Morgan and Jerry Doak.

Interestingly, one of the long-time Wickwire clients on the Babine was John Alevras, who later authored a book (*Leaves from a Steelheader's Diary*) that spoke mainly to his Babine experiences. Alevras gives a glowing account of his close friend Al Makkai, who was the ringleader of the group (the Babine Raiders) that he fished with annually on the Babine. Makkai and the Raiders were the six clients present when Oregon Bob and I spent those two evenings at Silver Hilton. I clearly remember Makkai holding forth and entertaining all within earshot, precisely as Alevras characterized him. Given the bond between the two, I now wonder if Alevras was also one of that six. His book references the buck bug twice but only in passing. Then again, he never mentions encountering anglers who weren't clients of the Babine lodges over almost 30 years of his presence there. That would seem to be a bit of an omission.

One more sidebar on the buck bug. Years after I began swimming that pattern in Skeena waters, I learned more about its development. According to New Brunswick lore, the buck bug was an adaptation of a relatively obscure 1950s fly known as the "Shady Lady." Emerson Underhill, a fly-tying institution on the Miramichi, found one of these among the collection of fellow fly tier Tom Balash at a farmer's market in Doaktown in the mid-1980s. Underhill thought the fly had possibilities, but not with its original black dun wool or black chenille body. Spun black deer hair was substituted and the buck bug was born. It quickly caught the attention of fish and fishers on the Miramichi. Almost three decades later, Underhill, who has tied many thousands of flies for Miramichi devotees far and wide, is pleased to observe that the buck bug has never diminished in effectiveness or popularity. In retrospect, that first batch I ordered from Jerry Doak was very early on in the history of the fly – more evidence it was an unknown pattern in the steelhead world before 1990.

There were too many memorable moments on the Babine to begin to cover in a few pages here. Heaven knows there are several other books out there romanticizing days spent and fish caught, which speaks to the river's reputation well enough. Still, I have one more fish story I like to remember.

I said earlier that my long association with Babine steelhead does not place them high on my list of performers. No doubt that is related to my fishing mostly the upper reaches of the system, where the fish are worked hard, with many captured more than once or twice. From a sample size larger than for any other river I have ever fished I can remember only two exceptional fish. Both required pursuit by boat to save fly lines and only one was landed. Both were pinned near the downstream limit of the beats frequented by the two upper river lodges, so neither would necessarily be comparable to the best of their cousins that were subjected to the far more intensive fishing and multiple captures awaiting them farther upstream. The landed fish was one of a kind.

Steve Pettit and I were camped 13 miles downriver. We chose that area for good reason: the water was at optimal level and we had two of our favourite pieces immediately at hand. Steve liked the Beaver Flats run proper while I preferred the run above. An island separates the two and divides the flow into roughly equal volume. I dropped Steve below the island and returned to the piece above. Those who fish that run want to do it from Challenge Rock, the large rock on river left reachable only by boat. I always did it the hard way, by anchoring my boat near the tailout of the run on river right and pushing my way upstream as far as water depth and velocity would allow me to wade. If I made my longest cast, and mended line repeatedly, I could wake a fly across the best of the holding water below that couldn't be properly fished from the rock above me on river left. It worked this day, as it usually did, and I managed a couple of fish from the smooth water just above the island. The side channel on river right was always worth a cast after the best of the rest was done. It isn't big water and the 14 ft. double hander I used that day was not the perfect tool. Nonetheless, I started by flipping a short cast or two, tenkara style, along the edge of the top corner. To my surprise, a large, rosy-cheeked head emerged to engulf a buck bug literally a rod length in front of me.

The struggle was typical and predictable. In and out, up and down, never more than about 30 feet from the rod tip. I stood my ground successfully for the first few minutes and at one point I was reaching for the leader to bring the fish to hand. That seemed to invigorate it enough that the line began fading from my reel in short but steady bursts. The gradient of the side channel picks up considerably as it moves down along the side of the island leading to the run below, where Steve was fishing. It took only a few minutes for my fish to be past the end of the junction of the two side channels at the tail of the island and headed for Steve, who was a couple of hundred feet below and blissfully unaware of what was happening. I scrambled down to the tail of the island, losing backing all the way, and shouted at Steve. He turned immediately and seemed to grasp my circumstances but then shouted back excitedly that he had just hooked a huge fish of his own.

I was now down to the arbor on my reel with nowhere to go. In that year, wading the side channel at the tail of the island would have been dangerous and wet and left me with no prospect for retrieving my boat. I searched the beach quickly and spotted a small log parallel to the water's edge that looked to be adaptable for a rod holder. I laid my rod on top of the log with the reel wedged firmly in the jagged butt at the upstream end. The pressure of all the line pulling straight downstream out there in the middle of the river below me was all that was needed to hold the rod in place. By this time there was nothing but the arbor knot left on my reel. Up the side channel I went as fast as I could manage and into the water above to push my way to my boat and haul myself over the stern. Down the river left side channel and around the tip of the island I went to retrieve my rod. Meanwhile, Steve figured he was into a monster fish of his own and unable to assist. I was happy to see my arbor knot intact and surprised to feel weight out there at the end of the line when I leaned back on the rod. It didn't take long to discover that my fish had gone under Steve's line and he had hooked my line about halfway between me and the fish. He was unknowingly playing my fish all the time I was retrieving the boat. Had it not been for the added pressure Steve was putting on that fish, I'm convinced we never would have seen it.

After freeing Steve's line I started retrieving my own, fully expecting the fish to be out of gas. That went as anticipated until the fish was once again close at hand. Off he went one more time, only to finally be subdued in a last-stand tug of war at the very end of the Beaver Flats run. Not until that fish was at my feet did we realize it was hooked near the top rear margin of its left operculum. That explained the difficulty in landing it but not how the hook had impaled itself there. I had seen every detail of the grab and I'm certain that fish was originally hooked in the right side of its mouth. Somehow the hook must have come free and lodged on the opposite side, well back of the hinge. Had the hook remained anywhere near its original position, that fish would have been landed as uneventfully as any other. A few inches' difference was worth a full 45 minutes of drama and as much adrenalin as any fish I've ever caught. Oh, and did I mention it just made the magical 40 in. mark?

Finally, there was "the year." In 1998, David Anderson, then the minister of fisheries, made the decision to close all coastal fisheries that were impacting rapidly diminishing coho salmon stocks. For the Skeena system that meant the usual commercial fishery harvest of roughly half the upriver bound steelhead didn't happen. Twice as many steelhead as expected is obviously good news, but there was more that added to what might reasonably be called the perfect storm.

Any given fishing season has multiple variables that influence fishing success. The number of fish is obviously one, but others – such as water conditions, weather, timing and the presence or absence of competition – can be equally significant. For example, Skeena country is well known for rainfall events that render every destination fishery pointless no matter how many fish might have returned to the system. The 1998 season on the Babine was one for the ages. The number of steelhead estimated to have entered the Skeena system was the highest ever in the 42-year period of record keeping to that point in time (it still is). Water conditions couldn't have been better, nor could the weather, and the competition from non-guided anglers hadn't yet reached intolerable levels. To add to all of that, the fish that returned that year included many displaying an unusually high condition factor (heavier than average for their length). Smolt survival and growth at sea are interrelated, so the abundance and size observations suggested strongly that the ocean environment frequented by

Skeena-origin fish of that cohort was as good as it had been in recent decades. The fact that those fish returned in a year when there was no commercial fishery was purely a bonus. I said at the time that a year like that would never come again. Steelhead guru Lani Waller and I discussed this point on several occasions in 1998 and later years. Given Lani's long association with Silver Hilton and all its catch records, he knew what I was talking about. We agreed we'd seen the best we ever would. The years since have strengthened that view. I often wonder how many others who were there in 1998 have any appreciation for what they enjoyed.

One could question my motives for returning to the Babine so many times over my fishing career. There wasn't anything special about its fish, other than the realistic chance of something exceptionally large, and it doesn't rank as a very scenic river in the many miles I could access below the fisheries weir. The greatest attraction for me was being able to leave the world behind. More often than not, I was fortunate enough to experience days when only my companion and I were on the water we chose to fish. Knowing there is good water ahead and undisturbed fish is worth something, as are streamside campfires, grizzly bears working the misty river edges at first light, unsuspecting moose and wolves, man's best friend beside me and the only sound the therapeutic babbling brook. I wish it would have lasted, but as with all things in this world, nothing stays the same. The river was and is seriously oversubscribed with guiding. That was acceptable in the 1970s and 1980s but the rod-day grab by guides pursuant to the implementation of the classified waters era in 1990 was the harbinger of change. Add to that the steady increase in Internet-inspired newbies in rafts and jet boats and the experiences once enjoyed can only be relived in the pages of my old diaries and photo collections. More people competing for the same water and fish is never a recipe for good behaviour, especially by some who try to make a living off it. Bryan Hebden, my 2012 trip partner and another retired biologist, and I endured some disgusting performances by one of the guides and his assistants. It was enough to cause us to consider whether the fishing experience had deteriorated to the point of no return. Decision made, I decided to have one last go at a favoured tailout that evening. It rewarded me with a beautiful 15 lb. male on my dry line and bug combination. Around our campfire that evening I told Bryan that might have been my last cast on the Babine. Several people, Bryan and myself among them, filed formal complaints about incidents with guides that year. Those responsible for "managing" the guides they license did nothing, despite several follow-up messages.

Example of large steelhead caught by the author on the buck bug fly on the Babine River.

10

UNHERALDED TREASURE
AND UNINTENDED CONSEQUENCES

If one could conduct a survey of all the licensed steelhead fishermen who reside in this province and ask them which of our river systems supports the greatest number of wild steelhead, I'll wager the vast majority of them would get it wrong. It's a safe assumption that there are more steelhead anglers from outside British Columbia who know the correct answer – the Bulkley/Morice – than there are from within. This simply reflects the fact that most of our resident steelheaders live in or near the southwest corner of the province. The centre of their steelhead universe is the Chilliwack/Vedder system and the Thompson its outer limit.

The Bulkley/Morice system is the largest and most accessible sub-basin of the Skeena watershed. It supports an average of 40 per cent of the total annual steelhead return to the Skeena system. The only highway across the central interior of the province parallels the Bulkley, and a large percentage of the population in the region occupies the communities scattered along its length. It isn't the Babine or Sustut. One doesn't have to be one of the Forbes 500 to afford to go there. You don't have to listen to banjos or endure the Hatfield and McCoy legacy of the Kispiox to partake of its bounty. A helicopter or jet boat is not a prerequisite (or should I say it didn't used to be?). You can spend an hour or two before or after work and expect to have as good a chance as anyone to catch a steelhead. Weekends were there waiting for you. It is (or, as will become obvious later, *was*) everyman's river: easily 150 miles of mainstem river set in a spectacular backdrop of mountains and aspen forests that light up the valley in steelhead season. Wildlife including grizzlies, black bears and even the occasional Kermode, elk, deer and moose –all presented potential viewing opportunities on a given day. In spite of all this, the river's virtues have never received the attention accorded all those other storied rivers of Skeena country. John Fennelly wrote extensively on his 1950s experiences on the upper Morice River and sparingly on the Morice near its confluence with the Bulkley, but the big river downstream got little more than passing mention. In the grand scheme of things, that probably helped keep it out of the spotlight for a few years.

I too gave little attention to the Bulkley before moving to Smithers in 1986. I recall a summer when, as a university student, I worked on the Canadian National Railway passenger trains between Jasper and Prince Rupert. I got to know a conductor and a trainman on those trips. Both lived in Smithers and both were dedicated Bulkley fishermen. Perhaps I should have paid more attention to their tales. All I can remember is how beautiful the river looked in late summer and how devoid of fishermen it always was. In 1972 I dabbled in the Morice for a day or two while waiting out fog that prevented a work-related trip to the fabled Sustut/Johanson area, and in 1979 I spent a few days of professional time there and on the Suskwa with a colleague who had undertaken one of the first radio telemetry investigations on Skeena steelhead. I also remember another angling acquaintance who worked at the Pacific Biological Station in Nanaimo in the mid-1970s. He had some fascinating tales of his Prince George days when he and his father made annual trips to the "Bear" (i.e., Suskwa) River before he had to return to classes at the University of Victoria. They killed a lot of big steelhead in late August of those years. But it was not until we bought our house overlooking the river that I began to appreciate the Bulkley's virtues. Given that I had visited the Babine for a decade prior to that, I often ask myself why I never bothered to spend more time on the river I had to drive by to get there.

The Bulkley near its confluence with the Morice was a well-known area that attracted a following of the original icons of the BC steelhead fly fishing community as early as the late 1950s but more so through the 1960s and 1970s. Many of them stopped off there on their way to and from the much more publicized Kispiox. Their successors still gather in the area even today, but the major focus has shifted downstream. This pattern is not unlike the one observed on so many other streams I have come to know. Obviously steelhead move upstream, so it follows that the easiest fish will be those never subjected to fishing pressure. Being first in line downstream is only logical. Why it took so long for that simple fact to manifest itself is a mystery, but it wasn't a bad thing for those of us who tuned in sooner rather than later.

My first serious efforts fishing the Bulkley followed the family move north in 1986. We didn't arrive until November 1, so what remained of that year's fishing didn't match the priorities of organizing a new house, getting kids set up in schools and becoming fully immersed in the contorted politics of Skeena steelhead.

Besides, I'd already spent a full week on the Sustut and a rain-shortened one on the Babine earlier that fall. The following year was the first real opportunity I had to begin to develop my awareness of the river and put that knowledge into the context of all I had come to know elsewhere. Only time can do that.

Having a fetish for big fish, I liked the idea of the Suskwa area as a good place to jump-start the Bulkley learning curve. That earlier radio telemetry program I spoke of revealed, among other things, that some of those big Suskwa-bound fish spent a lot of time in the Bulkley, sometimes well upstream from their eventual destination. That was worth some follow-up.

One of my earliest excursions to the Bulkley/Suskwa confluence area was on Labour Day weekend in 1987. It was a family outing. With Lori and our three kids, aged 4, 6 and 8, toodling down the river didn't make for much in the way of serious fishing but that wasn't what the day was supposed to be about. The family lounged on the beach on a beautiful sunny day while I threw a few casts with a float-fishing outfit nearby. The first fish of the day came easily. There was nothing to suggest there was anything special about it. It didn't fight long or hard, there was no aerial display and I don't think it ever got more than perhaps 60 feet away from where I'd hooked it. The water was clouded with the Telkwa glacial contribution associated with the recent hot weather so I didn't get a clear look at the fish until the very late stages. My response when I did see it? "Lori, get the net!" That was my one and only 30 lb. steelhead. No estimates here. I had my custom weigh bag and precision scale in the boat. The scale dipped to a hair over 30, as confirmed by Lori. A couple of quick pictures and off he went. Not a bad start.[21]

The author's personal best 30 lb. steelhead, caught at the Bulkley/Suskwa confluence on September 6, 1987.

Bulkley viewscapes.

There were several other trips to the Suskwa confluence area that year and several more large fish whose pictures are carefully stored in my personal collections. Suffice to say it was not hard to fall in love with that part of the Bulkley when conditions were right. One key point is worth adding to the accounts of the year. My diary notes that in the six days spent in the Suskwa area between September 6 and October 7, 1987, I never encountered another boat or even another angler. I'll come back to how that changed over time.

The next year, 1988, was memorable for its contribution to my "nine lives" stories. Friends Steve Pettit and dry fly guru Bill McMillan had journeyed north to join us in Smithers for some Bulkley time. Steve had his drift boat in tow and the plan was that he and Bill, using our home as a base, would do overnight drifts from Smithers to Trout Creek. Government workers had been on strike for several days prior to their arrival and I was on picket duty at my workplace every second day. That left me with more downtime than anticipated but limited flexibility as to how I could use it. The Skeena was at a dream level I'd never seen before, and I had enjoyed some off-the-charts fishing on several of my non-picket-duty days in the week prior to Steve and Bill's arrival. It seemed worth a day trip to the big water near the mouth of the Kispiox to try to get the visitors into some of the bounty before they began their Bulkley drifts. My diary reminds me that Bill took a magnificent 18 lb. hen on a dry line that day. Back to Smithers we went, me for picket duty the next day, they to launch at Chicken Creek, five minutes from my door.

As fate would have it the dry spell and premium water level didn't last. A couple of days later, Steve and Bill got caught in what proved to be one of the nastiest storms I saw in all my days in Smithers. A major front moved in from the north coast, up Douglas Channel, across the top of the Copper and Telkwa watersheds and straight on to Morice Lake. The lake and river system rose sharply overnight on September 29. Twenty-four hours later, the entire Bulkley system was experiencing major high water. The storm was so intense and the rise in water so rapid that even Morice Lake was stirred up to the point that water clarity precluded decent steelhead fishing for weeks afterward. Steve and Bill managed to get to the Trout Creek takeout just in time to avoid the major pulse and debris working down the system behind them.

Water clarity being what it was, all through October and most of November that year I never fished a single day. On the last Sunday of November (the 29th) I was gazing out our kitchen window at the river below while washing breakfast dishes. The thermometer at the window's edge registered something like 9°C, the river had dropped with a recent bit of cold, dry weather and the morning sun highlighted blue water instead of the persistent grey. I thought to myself, "Why not?" Within an hour I was in the water at Chicken Creek. My jet sled had already been mothballed for the year so I chose to drift from there to Trout Creek in my trusty little Zodiac Youyou. That is a skookum little 7 ft. double-ended inflatable I'd used countless times on some pretty wild water over the years, so I had no reservations about my plans for the day. Lori and the kids would meet me at Trout Creek at dark. What could be simpler?

What I didn't pay enough attention to was how long it would take to make the 18-mile drift. Steve and Bill had done their overnight thing leisurely, fishing all the way through. I figured I would minimize my fishing in the water I already knew over the first few miles and spend what fishing time there was on the water they had done best on well below. Fishing wasn't top of mind anyway. Learning the river and what it offered for future days was my main motive. I thought I knew about Driftwood Canyon and a second canyon piece downstream (Reiseter), but neither had been highlighted as particularly treacherous. I was unconcerned.

As I passed Driftwood Creek and the mini-canyon that began immediately below its confluence, I figured round one was already behind me. A few hundred yards below, I discovered otherwise. The real Driftwood Canyon was right there in front of me and the whitewater I was headed for was too big for my little Zodiac. Unfamiliar with how to get past this point I decided to go ashore and scout things from the beach. One hard pull on the oars to properly orient myself to get to river right proved disastrous. The rowing frame, which I had confidently crafted out of PVC pipe while sitting around home thinking about the fishing I wasn't getting in during previous weeks, snapped at the point where the starboard oar had been mounted. Drifting helplessly toward the worst of the whitewater 30 or 40 yards distant, I frantically ripped the oar from the broken frame and used it to draw the little raft within wading depth of the beach. I jumped out and dragged myself to high ground. A quick assessment told me to rid the raft of its frame, lash the two oars together with the bow line of the Zodiac and paddle, kayak style, to river left where it appeared I could portage the ugliest piece of water. I put my fishing vest on the floor of the boat, broke down my rod and cinched the wading belt on my neoprene waders as tight as I could.

The improvised kayak paddle worked remarkably well. I crossed the river easily and portaged the raft across the bedrock ledge bordering the cascade on river left. Back to paddling I went. It was pushing 3:00 p.m. by the time I cleared Driftwood Canyon. Still less than halfway to Trout Creek, with the rest of the drift through water I had never seen before, this wasn't an inviting prospect. Unlike today, when cabins and houses are spotted at regular intervals, no sign of human habitation was visible from the river anywhere on my flight path. The only way out was down the river. No communication, rapidly approaching darkness, declining temperature – this was the stuff you hear about on newscasts after they find the body. I kept wondering what Lori and the kids would do when I didn't show up at the appointed time.

One's senses are sharpened under such circumstances. As darkness descended I became acutely aware of the river's sounds. There was not enough light from a late-rising half moon in a now-cloudy sky to detect turbulence ahead, so listening became critical. I felt my way through Reiseter able to see only the white of the swirls and tops of the standing waves within my reach. After that I just paddled as hard as I could manage to get my tardy ass to the rendezvous at Trout Creek. Lights appeared before I expected them. As I stood and straightened my cramped legs, Lori caught my movement and was there in seconds. First hugs, then a lecture. I think it was around 7:30 p.m. She had walked from the truck, parked near river's edge, back across the highway to the Trout Creek Store and called my colleague, Mike Lough, in Smithers at about 6:00 p.m. On hearing the circumstances, Mike had advised her that I couldn't make the trip in the time expected, and so she should wait another hour or so and then call if I still hadn't shown. She was about to make that second call when she saw me. I promised I would never put her in such a stressful situation again. The warm cab of the truck and the banter of the kids were mighty welcome on the ride home.

The next several years saw many days spent on the section of river I had drifted on that eventful November day. It became standard for me to launch either at Trout Creek and run upstream for the day or at Chicken Creek and run downstream. Either way, there was the promise of a lot of prime water that hadn't been touched on the day of arrival. The only guide with any regular presence on that section of river operated drift boats that did overnight trips between Smithers and Trout Creek with a tent camp near Reiseter Creek serving as the accommodation. One always knew where and when the drift boats were likely to be encountered; there was no difficulty avoiding them and minimizing overlap such that everyone was able to find unmolested fish. The river between Houston and Smithers was always busier because it was much more accessible and easily navigable. Those areas were also fished more heavily by guides. All the more reason for my choice of water below Smithers.

The rapids, Driftwood Canyon, Bulkley River.

Learning the nuances of the sections of the Bulkley I liked best was a voyage of discovery. No one ever led me down the river or pointed to where I should fish. I took great pride in finding my own fish and learning how to make them respond. During that process I discovered a spot that eventually proved to be a very good barometer of what was in the river. It was a small piece, the actual holding area about the same size as my boat. Bedrock defined it, so there was never any change from year to year.

That special place revealed itself on a September day in 1993. The river was low at the time and I was exploring offbeat places that might hold promise. On idling through the area I noticed a break in the current not unlike the one I described earlier at the tail end of the gravel pit run on the Nanaimo. Closer inspection revealed that the break was created by a bedrock ledge that fell away into a cobble-lined pocket below. Halfway back was one large rock that was barely detectable unless you were closer to it than you should have been to fish effectively. The first time waking a bomber behind the rock produced a rise for my son, Brock, but no comeback. I made a mental note.

Several days later, I was fortunate enough to be able to spend some time with long-time friend, respected angler and conservationist Pete Broomhall and his trip partner, John Taylor. We fished prime water well downstream that day and managed a couple of fish in the usual places. On the return upstream I swung the boat across and hovered near my new spot long enough to see a

The first fish landed from the author's secret spot on the Bulkley River, September 23, 1993, and another caught by the author's son on September 6, 1998.

fish resting behind the rock. The river was so low, and that fish so exposed right there in front of us, it seemed improbable that we hadn't already spooked it before we could show it a fly. Undeterred, Pete made a delicate cast to it with one of his Art Lingren–tied Black Practitioners. The fish climbed on immediately. That was the first of an amazing number of fish that came from that one little bathtub for days and seasons after. The single best visit on a warm September afternoon a couple of years later produced five fish on a bomber while Lori and our two daughters basked on the front of the boat in bathing suits. After I returned from walking the third or fourth fish to the beach and releasing it, I began asking my audience if they had any appreciation for what they were witnessing. Five fish hooked on a dry fly, all landed, all in the space of an hour and a half! Oh well, I knew Brock would get it when I had a chance to share the story with him that evening.

My special spot remained a closely guarded secret for many years. If there was any possibility that someone was observing me, I wouldn't stop there. Early and late in the day worked best. It was rare not to find a fish there if the water height was anywhere near suitable. Today I refuse to fish it because there is too much boat traffic racing up and down the river to avoid being seen. Let those others find their own fish.

There is one more bit of personal Bulkley history I like to share. I had the unfortunate experience of losing some cherished rods and reels in an accident involving a defective swivel on a helicopter longline slinging our gear up the Dean River on September 9, 1997. The pilot's insurance eventually covered replacement of the rods and reels – not with the originals (a Hardy Perfect and a Hardy Dural St. John) but with a couple of Sage Spey rods and suitable reels. That was my introduction to double handers. All that remained was to obtain the appropriate lines and get on with another learning curve. Who better to aid in that process than Mr. Rio himself, Jim Vincent?

Jim and I had known each other since the earliest days of the classified waters regulations formulation processes. At the time he was the consummate steelhead bum and perceived as the devil personified by all those who didn't like non-residents, especially "aliens" (non-Canadians), cluttering up our rivers. It didn't help that he spent a lot of time on the Dean and the Kispiox and fished circles around most of his critics. I always admired his prowess as an angler and I respected his knowledge of circumstances on a lot of blue ribbon waters, as well as his penchant for telling it like it was. By the time I had a 14 ft. Sage in my hands, Jim's Rio fly lines dominated the steelhead fly fishing world, and we were well enough acquainted that I felt comfortable in seeking his advice. The stage was set for a couple of days together on the lower Bulkley in mid-October 1998: the year of the steelhead.

The interesting part of those days with Jim was the effectiveness of that silly little fly I had fallen in love with years earlier. Jim was much more conventional in his approach, preferring larger patterns and generally heavier gear. He was a big, strong guy wielding an equally powerful Thomas & Thomas rod and his own lines to beautifully deliver his fly of choice twice as far as my best cast. On a run we fished on each of the two days, I worked down behind him so I could observe and try to mimic his casting. Meanwhile, I landed six fish to his one, although he also lost a couple. My success obviously had nothing to do with my casting ability, nor with anything I was doing differently once the fly hit the water. My diary entry: "made a buck bug believer out of him."

My honeymoon with the Bulkley didn't last, though. The same can be said for every other iconic Skeena tributary. The Bulkley miles I most enjoyed remained less travelled than other parts of the river for most of the first decade following the implementation of the regulations that were supposed to preserve the values they offered. But, short years later, they converged to the same end as everywhere else. Too much will be lost if some effort isn't made to put the essentials of that process on record.

The door opening on the Bulkley began with the bonanza steelhead years of 1984, 1985 and 1986. Excellent catches catalyzed the growth of a guiding industry that had barely been detectable previously. The half-dozen small-time locals who were one-man shows subsidizing their fishing habit by guiding during days off from their regular jobs had never caused any waves. Then came the aggressive import who set up shop in Telkwa and promptly hired a half-dozen or more assistants to row his customers down the river every day of the week. It was the equivalent of the big box store arriving in Sleepy Hollow. The new show became a constant presence in the face of locals not accustomed to seeing any of it. It also had the effect of pushing others to play catch-up and try to get a piece of the growing action themselves. By 1986 the pressure was mounting on those of us in the management agency to "do something." A moratorium had already been imposed on the issuance of new guide licences, so the number of guides was not the issue. It was the sharply increasing effort of those with a foot already in the door that most concerned the locals.

Any potential for public dissatisfaction over government actions is dealt with by exhaustive consultative processes. Those leading to the regulatory prescriptions for how many guides and how many rod days would be allocated for the Bulkley (and every other classified water in the province) consumed three years. New regulations came into effect on April 1, 1990. The gold rush was over, or so most people thought – and it could have been, but for the indifference of the statutory authority, the person legally responsible for signing off on rod-day quotas. The man had never spent a day in his life on a classified water. I'm convinced he never owned a freshwater fishing licence, either. His lack of respect for those of us in-service who partook of such things was seldom disguised. Predictably, his trivialization of all that had been done to try to preserve a piece of the steelhead-fishing future for the resident angling community proved disastrous. A bit more explanation may help others to appreciate my opinion.

The fishing seasons of 1986, '87 and '88 were set as the reference period for guides and guided rod days on all the rivers that were designated as classified on April 1, 1990. On the Bulkley there had been seven guides in operation. The angling-guide reports filed by each of them as a condition of their licence served as the basis for defining rod-day allocations. The actual policy stated each guide would receive the average of the two highest years' totals for the three reference years. The problem was that the guide reports were oftentimes deplorably incomplete and inconsistent. I state this unequivocally because I was the person responsible for trying to make sense of them all and prepare recommendations for the statutory authority to approve. My efforts to explain both the deficiencies and the potential for allocating more days than were supported by any credible evidence fell on deaf ears. Warnings about the possible problems of inflated and unjustifiable rod-day quota allocations, once issued, were dismissed on the premise that any days not used could and would be clawed back through the powers available under the new regulations. In the end my instructions were to determine the maximum number of days each guide had used in any of the three reference years. These figures subsequently became the rod-day allocations defined in law. Seven guides and 1,504 rod days were the figures as of 1990. They remain in effect today.

The measure of how inflated the rod-day allocations were became abundantly clear in the first three years following enactment. In year one, the seven Bulkley guides reported using 64 per cent of the number of days allocated. The next year it was 27 per cent, and the year after, 36 per cent. Some of those numbers were highly doubtful because the largest single rod-day-quota holder failed to produce evidence of licence revenue remittance or licence counter-foils to match his guide reports. There were no consequences. And so it went.

Five years later the real estate agent, the train engineer and two other small-time locals that no one I could find had ever witnessed guiding on the river had sold their rod-day quotas to newcomers who became very obvious players on the scene. Not a single unused day was ever clawed back by the statutory authority. They authorized the sale and transfer of every one of them. The fly fishing school operator whose original authorization (a permit, not a licence) was supposed to be for a single eight-day block, and only to teach fly fishing, had become a full-fledged guide complete with assistant guides in jet boats. Shortly after, the second-largest operator – whose original quota ranked as, relatively, the least supportable of all – sold to outsiders who quickly developed his little homestead into a full-fledged operation that put all those original ficti-tious days to use. Today there remains only one of the original seven rod-day-quota holders, and that operation has been akin to a rod-day brokerage that, for years, allowed an assistant guide who advertised openly on the Internet as a guide with 30 years' experience on the Bulkley to run his own business. He even had a foreign booking agent broadly advertising his operation. The 1,504 rod days that were never anywhere near used in the years immediately follow-ing their allocation have been all but consumed completely in recent years. In addition, several operators now guide in late August and/or November when there are no limits on days used. So, even the so-called shoulder seasons, when local residents could for once enjoy a piece of river without having to compete with the professionals, were compromised.

There is more. In the mid-1980s, when guiding was emerging as the issue of the day on the Bulkley system and elsewhere, there was only one guide who employed assistant guides. Those assistants worked from drift boats, not jet sleds. During the 2014 season, the seven Bulkley guides employed 41 assistant guides. Almost all of them operate from a jet boat whenever they are on the water. Some of the 41 may not frequent the Bulkley because their employer also has other operations on other rivers. However, that could be offset by the number of assistant angling guides who were originally endorsed elsewhere, by guides who don't work the Bulkley. Any assistant guide, once licensed, can freelance for any guide. If one of them had worked, for example, the Dean River earlier in the year and then requested a Bulkley guide to endorse his licence so he could work that river too, no record would ever find its way to the licensing agency. The point is that an assistant guide is, functionally, no different than a guide. Each represents another boat, more anglers and less opportunity for everyone else seeking to tap into a fish supply that never gets any larger and a space that is fixed. The proliferation of assistant guides allows the rod-day-quota holders to concentrate use on all the best weeks of the season. There is no limit on how many clients a guide operation can squire around in jet boats on any given day. Do people understand that assistant guides who fish right along with their clients (and they commonly do) aren't counted against the rod-day quota of the guide employing them? I'm defined as a rod-day regardless of whether I fish an hour or an entire day. Why shouldn't the same accounting be applied to guides and their assistants?

The number of guide-operated lodges and camps spread along the Bulkley and Morice is yet another festering sore point. Most are now foreign-owned. Hearkening back to the period that was supposed to serve as the benchmark for guiding, there was one base camp on the Bulkley and one on the upper Morice. The upper Morice camp is relatively unchanged from that period other than in who owns it. Not so on the Bulkley. The base of operations for the original Bulkley operation was the quaint old Creamery in Telkwa. Today it is a spacious, deluxe new facility a few miles upstream. The angling guide licence holder in charge of that facility is now also in command of the original upper Morice River lodge plus three more satellite camps on the lower Bulkley River. Nearby on the other side of the river the homestead that was being upgraded just as its owner's grossly inflated rod-day quota was being settled blossomed into another full service lodge. More recently, one of the longer-serving Bulkley guides sold his modest rod-day quota to a foreign-based owner of multiple operations on the Skeena system and elsewhere. Clients of the original guide were accommodated in local hotels. Now they reside in privately owned riparian facilities well downstream from Smithers. That operation commenced in 2015. Two guides based in the Kispiox Valley also frequent the lower Bulkley in the Suskwa area.

Finally, there is the ever-increasing business of "hosted trips." It is now common to find advertisements by well-known sport fishing industry personalities from the United States (and others, less well known, from Europe) offering up week-long trips to the Bulkley or any number of other rivers. Some of them own fly shops; many guide on waters other than ours. Several are booking agents for an array of renowned international fishing destinations. The people promoting these trips get a free ride for bringing in a week's worth of clients for the rod-day-quota holder. There are also the filmmakers, rod makers, fly tiers, authors, casting clinic facilitators, celebrities etc. who now show up regularly. No one ever checks on any of this to judge the legalities involved. When one of these high-profile people is running a casting clinic or promoting the sale of custom-made rods or making another fly-fishing video, who counts as the guide or assistant guide? Who counts as a rod day used? Who is on record as the guide when the fly guy from California arrives with his half-dozen clients in tow? When the film crew takes over a camp or prime water for several days to shoot self-promotional television shows, who actually buys a classified waters licence and what goes on the books as rod days used?

The underlying policy for the classified waters initiative was preservation of quality angling for resident anglers. The level of commercialism that has evolved since implementation was precisely what the classified waters initiative was intended to prevent. Too many senior government people who refuse to educate themselves still contend that the commercial traffic on the Bulkley is substantially unaltered over time. They succumbed to the guides' demands to beat up on the do-it-yourself boogeymen from Alberta and the evil aliens from outside Canada, despite a dearth of any credible evidence in support of their claims that those two groups were the source of alleged crowding. The ultimate eye poke was the orchestration of a regulation (2013) forbidding non-Canadians from fishing the Bulkley on weekends – unless they were guided. Whereas Saturday was traditionally the changeover day for guides and the day the locals could look forward to for relief from the flotilla of guide-operated jet boats, the guides promptly shifted to weekdays to rotate their clients. What better deal could they possibly have received courtesy of the regulators? What was once everyman's river, to be protected largely for the benefit of resident British Columbians, is everything but. The best line I've heard to capture the Bulkley scenario today is this: "When is a number not a number?" The answer: "When it's a rod-day quota."

At the risk of boring all whose eyes find these pages, the book of Bulkley needs to cover one more subject: Moricetown and the ongoing steelhead-tagging program that is supposed to provide ironclad data on the number of steelhead ascending the river beyond that point. In my earlier book on the history of Skeena steelhead, this subject was addressed in a level of detail that won't be repeated here. A point or two should be emphasized, though.

It isn't the functional-level Wet'suwet'en people who do the capture and tagging of steelhead at Moricetown who need to be called to account. They are paid to do what they have been led to believe is good science that will greatly enhance the knowledge base on Bulkley-system steelhead and position band members to effectively manage fisheries resources in the future. DFO websites label these initiatives as "capacity building." Instead, the problem rests with the people who have never clarified what that is or how it applies to Moricetown. Twenty years and millions of dollars later, the only detectable benefit is seasonal job creation for a handful of people. I've asked this question repeatedly: What value can there be for an estimate of the number of steelhead passing Moricetown when that number doesn't materialize until many months after the end of every fishery that influences it?[22]

The most recent justification for the ongoing Moricetown exercise is that "we" need it to calibrate the test fishery that DFO conducts in-season near the mouth of the Skeena River. DNA analyses of fish taken there reveal what proportion of the total season catch originates from the Bulkley system beyond Moricetown. In theory, if there is an accurate estimate of how many fish get to Moricetown, their proportion in the test fishery can be translated to an actual number and the estimate of the total number of steelhead entering the Skeena can be refined. All good – except the Moricetown procedures seriously overestimate the Bulkley/Morice population size. My Skeena book explains all of this in greater detail. The bottom line is that inflated steelhead-population estimates become the figure against which estimates of the harvest rate of steelhead in the commercial fishery are developed. The new math is then used by DFO and the commercial fishing industry to strengthen their position that river-mouth nets have never been an issue in steelhead conservation.

All politics and pseudo-science aside, it is the impact of the Moricetown activity on the fish themselves that most concerns me. The ever-increasing capture efficiency of the steelhead taggers is not fish friendly. The scientific literature on the subject of harm done to steelhead and other closely related species by beach seining, dip netting, tagging, keeping fish out of water for extended periods etc., clearly documents the negative consequences. Suffice to say, nothing is more damaging to a steelhead than a large mesh dip net (effectively a small mesh gillnet) used to capture it and pack it airborne across catwalks and along canyon trails while it thrashes itself motionless before being dumped into a large plastic bucket for "processing." We're not talking small numbers here. The worst of it was in 2010 when the cumulative total catch by beach seiners and dip netters was just short of 10,000 steelhead.

If you're an angler who has spent years fishing the Bulkley and encountering survivors of these processes you know a few things. First, virtually every tagged steelhead caught is lethargic and bearing evidence of wounds and abrasions that call into question that fish's ability to survive, reach its intended destination and/or reproduce successfully. Second, you will have caught many more fish that were tagged by beach seiners than by dip netters even though the number of steelhead dip netted and tagged far exceeds the number of those tagged by beach seiners. Third, you will be familiar with the recovery of dead and dying tagged steelhead many miles downstream from Moricetown, and even in the Skeena. Finally, you may have noticed the ratio of tagged to untagged steelhead has increased steadily over the past decade.

One criticizes anything to do with First Nations fisheries at their peril in the world of the present. The socio-political pendulum is hard over on the side of First Nations and governments have obviously climbed on that train. Court decisions are the foundation. Operational-level fisheries agency staff are no longer in a position to reveal readily available data that underscores the damage done to steelhead at Moricetown. Even guides whose sample size dwarfs the cumulative total catch of half the Bulkley anglers do nothing to help their own cause. The major players there decided long ago not to report any information on tagged fish. In fact, they freely confess to ripping tags out of fish so their clients don't know the fish they just landed has previously been touched by human hands. No data = no problem. Without proof of harm the program carries on, funded by what amounts to your tax dollars and mine.

Placing the Wet'suwet'en people at Moricetown in the crosshairs gets us nowhere. What the situation screams for is a longer look at the costs and benefits of fish and fishing. Someone in a position of authority needs to step forward and do the analyses. Surely the biggest wild steelhead fishery in British Columbia is worth that. Surely there are reasonable people in all three levels of government that can devise more responsible ways for the Wet'suwet'en to achieve their fisheries management goals.

A series of seven photos of the Moricetown steelhead mark and recapture populations estimation methods and procedures, Bulkley River, September 10, 2010.

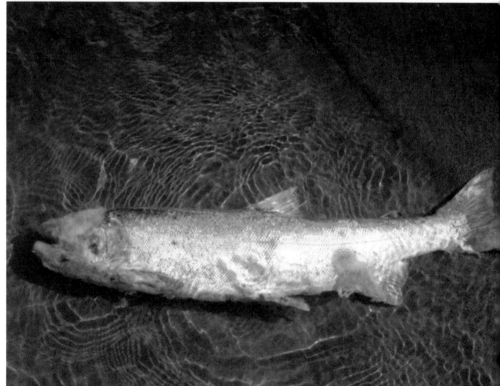

Examples of steelhead bearing tags applied at Moricetown and later discovered
at various points along the Bulkley River

11

MANY ARTERIES, ONE HEART

The old guard of the steelhead advocacy community recognizes Jim Culp as one of its most passionate and dedicated personalities. I know Jim from countless interactions over at least 40 years but if I had to recall a single moment that crystallizes my memories of him it would be a comment he made about the Skeena River. I believe it was in one of the many sessions dealing with the future of fishing on all those magnificent rivers in the Terrace vicinity – the Copper, the Lakelse, the Kalum. I think it was in the late 1980s, when the classified waters regulations were being formulated. All eyes and minds were on the tributaries when Jim admonished us, saying that none of them was as important as what he referred to as "Mother Skeena." Those words have stuck with me just as much as Tom Morgan's line about little fish in big bodies.

Of all the rivers I have known and fished, the Skeena comes closest to illustrating that overworked word – awesome. How many times did I travel Highway 16 from Smithers to Prince Rupert to deal with steelhead issues over those 13 years in the Skeena management trenches? How many trips on the water from Kispiox to Kitwanga, from Exchamsiks to Kitselas? The journey along the Skeena was as captivating the last time as it was the first. In its seasonal moods and with its stupendous backdrops of mountains and tantalizing views of sub-basins carving their way through them, there was never a time I wasn't gripped by the majesty of it all. No other river in British Columbia comes close to providing the opportunity to pursue steelhead in such spectacular surroundings, from headwaters to tidewater, for so much of the year. Throw in the fact that the river's fish are 100 per cent wild and the largest on the planet, and the quest for perfection need go no further.

My early days on the Skeena proper came well after exploratory trips to several of its renowned tributaries. I'd spent some time around the Kitwanga confluence in 1979, but it wasn't until 1984 that the big river really captured my attention. That year, Oregon Bob and I had some wonderful fishing on the Babine, followed by much the same for a couple of days of drifting the Kispiox. Ron Tetreau, a long-time Kispiox Valley resident, as ardent a steelheader as ever there was and a fellow government employee at the time, had told me of the Kispiox/Skeena confluence area and suggested Bob and I might want to take a look at it on our way past. He said very few people ever bothered to fish that water.

September 30 found Bob and me at what is now known as the market garden, the road access point nearest the river below. We parked and searched for a way over the steep bank, eventually finding ourselves at water's edge. From there we hiked upstream to what looked to be a very large and inviting piece. I later dubbed that run "Obvious." We managed two fish landed and two more lost in a brief session before having to bid it goodbye and begin our journey south. My diary entry: "Appetite whetted for a return trip with Valco. Gorgeous country, fabulous water, not a soul in sight. I'll be back."

Bob and I did return in mid-September the following year and had decent results over a couple of days prior to another Babine adventure. Once again, we had all the prime runs in the vicinity of the Kispiox confluence to ourselves. I couldn't fit Skeena time around Sustut and Babine excursions in 1986, but moving north that year set the stage for serious days ahead.

By far, the Skeena sport fishery that attracted the greatest interest for most of its history was that of the huge chinook that originated in the Kalum system. Those trains I worked on back in the late 1960s crossed the Kalum near its confluence with the Skeena, and there was always much evidence of fishing activity both there and at a number of other nearby spots downstream. Not unlike the Kispiox when its steelhead hit the *Field & Stream* annual fishing contest leaderboards in the 1950s, the Kalum's giant chinook became an internationally renowned draw as soon as word of a then world record 92 lb. fish reached the airwaves. That fish was taken by Terrace resident Heinz Wichman on July 19, 1959. Steelhead were a welcome bycatch in the Skeena proper as the chinook runs tapered off and coho started, but there was no significant target fishery for them until years later.

The mid-1980s were the tipping point for the lower Skeena for all the same reasons experienced at places farther up the system. It was an interesting time to be introduced to the vocational and avocational sides of it all. Weekends and family camping trips to some of the magnificent bars on the river below Terrace were the stuff of lifelong memories. So too were the Monday-to-Friday dealings with the blossoming guided sport fishery that was all but commandeering many of the best fishing locations by setting up semi-permanent camps that catered exclusively to European anglers. I know enough about the evolution of that phenomenon to offer it up as a candidate for reality television. Confidants explained that much of what transpired had roots in financial dealings and outstanding debts in a country nine time zones away. Few of the players I dealt with would ever be accused of high moral standards or of playing by the rules.

Politics aside, the lower Skeena bars provided some wonderful fishing opportunities, especially if one had a boat to get to less-travelled reaches. Bell-tipped rods were set in a fenceline of home-built holders while the little people tended campfires and tried to catch the ever-present juvenile fish scurrying around the shallows. A ringing bell brought instant debate about whose turn it was to do battle with whatever had eaten one of those hummingbird-sized Spin-n-Glos. We saw them all at one time or another – chinook, coho, pink, chum and the most beautiful steelhead one could ever wish to encounter.[23] So much for July and the front half of August for the upriver folk like us who journeyed to Terrace on weekends.

The author's father and children with a prime steelhead caught on one of the lower Skeena River bars in early August 1988.

Come September, fishing the mainstem Skeena took on a different aspect. No more shorts and T-shirts, no more lawn chairs, no more uncertainty as to what was ringing the bell. Now it was all about donning waders and searching for the leviathans bound for the Kispiox, the Babine and beyond. The Skeena below Terrace was emerging as more of a target fishery for September migrant steelhead by the late 1980s and into the '90s, so fish whose predecessors once passed through those waters unmolested faced a new and escalating fishing effort. Other than the area around the Kitwanga confluence, relatively little sport fishing occurred between the Copper River confluence and Anderson Flats at the mouth of the Bulkley. Mind you, those reaches were the epicentre of the First Nations harvest of steelhead. Set gillnets abounded through Kitseguecla and both set and drift nets were ever present around Kitwanga. The Hazelton through Kispiox area, especially around Glen Vowell (now Sik-e-Dakh), was also fished heavily by set and drift nets. That has not changed; in fact, nets are probably more prevalent now than ever before.

Fishing the Skeena in the Kispiox area in the late 1980s and into the '90s was high on my all-time list of quality angling experiences. The water had to be at a respectable height, of course, and there were extended periods when that was not the case. But, when the water and timing were right, there was

The decaying catch of steelhead removed from an unattended gillnet by a federal government fisheries officer, Skeena River, September 24, 2012.

nothing quite like it. The best for me was 1988, a year I referred to earlier. By mid-September the river level had dropped and the water clarity improved to the point that no time was wasted in searching the big runs for the travel lanes and stopover areas. Nature had set the table as if it had been ordered from the menu. Conditions just kept improving – until the same weather system that took out Morice Lake and everything downstream on September 28 and 29 descended on the Skeena as well.

The September fishing I found in 1988 will never be duplicated. The combination of a constantly replenished supply of large, frisky fish, perfect water conditions and zero competition was the benchmark against which I evaluated all that followed. The every-second-day fishing opportunity afforded by the government employees' strike and picket duty didn't help my monthly paycheck but the sacrifice was well worth it for the fishing that was had. My son, Brock, was 5 years old at the time and joined me on several of those days. His first ever steelhead came on one of them.

We'd launched the boat at the mouth of the Kispiox earlier that morning and fished some of the water below. Several fish later, we took a lunch break and ran up to the mouth of the Shegunia River, directly opposite and in full view of Kispiox Village. After soup and pan-toasted cinnamon buns, we pulled out a small plastic case containing a matched Daiwa rod and spinning reel. I had recently purchased the set so Brock could do a bit of fishing on his own. We assembled the little four-piece rod, mounted the reel and rigged it with a short leader weighted with a split shot and a small single hook decorated with a couple of jarred single eggs. He began making short casts right in front of us where the Shegunia fanned out and spilled over a broad, ankle-deep gravel bar into a fishy-looking trough as it met the Skeena. There was usually a Dolly or two hanging there, picking off dislodged pink salmon eggs rolling over the bar from that year's abundance of spawners, which hovered over every available piece of gravel just upstream in the Shegunia.

Mere minutes later, that little Daiwa developed a serious bend. No, it can't be, I thought. The odds on a little guy landing a 9 lb. fish on a rod best suited for sunfish were out there with the Toronto Maple Leafs winning the Stanley Cup, but that fish was destined to make its ultimate sacrifice. A few short bursts, a couple of jumps and, magically, there it was at our feet. Brock had seen more steelhead caught and released by the ripe old age of 5 than most anglers will in a decade. He took it for granted that we would do the same with this one. I explained that I had kept my first one and it was all right with me if he wanted to take this one home and show Mom. Such were the times.

The author's son with his first-ever steelhead, Skeena River, September 18, 1988.

By 1988 I'd seen as much as I needed of steelhead in British Columbia to appreciate that those two weeks on the Skeena in late September of that year were more than just rare. No competition, no pressure to move; time to experiment with gear, to break out the coffee pot whenever, drink in the scenery, do some photography, savour a fish – all this bordered on incomprehensible after 25 years on the rivers of southwestern British Columbia. Even the fishing I'd had in the Sustut and Babine years earlier didn't begin to compare with Skeena '88. Maybe I was just plain lucky to be able to enjoy those Skeena days in solitude. Surely there were others familiar with the opportunity available. After all, the world-renowned Kispiox was right there. Some of those early icons of the steelhead fly-fishing fraternity who had been fishing it for decades must have appreciated that low water periods, especially early in the season, would have Kispiox fish kegged up awaiting the rains as well as Babine, Sustut and other stocks moving through at the same time. Guides were certainly active up the Kispiox, but I encountered nary a one on the big water below. The Gitxsans next door in Glen Vowell fished steadily from late June through early August but were similarly absent that September. Even Bob York, the keenest and most aggressive of all the Kispiox devotees, was nowhere to be found in 1988. I never expected all that to last, and it didn't.

With my appetite having been whetted in 1988, I started much earlier the following year. Three trips in August were fishless. Labour Day weekend produced only one steelhead. Dirty water prevailed through those early weeks and didn't improve measurably through the remainder of September. It was glacial melt rather than rain that kept the Skeena's clarity at undesirable levels. A half-dozen trips saw only a fraction of the fish encountered a year earlier. The Kispiox was dead low through September 1989 and fishing poorly. The crowd that normally occupied all the prime water well up the valley began to trickle downstream. A couple of new wave fly guys started using brightly coloured egg patterns on split shot weighted leaders and strip casting to fish plainly visible in one particular run just below the bridge crossing that led into Kispiox Village. Their kiss-and-tell habit back in camp precipitated a pre-dawn race to the spot in subsequent days. That was followed by near fisticuffs over what constituted fly fishing. Aggrieved anglers went so far as to visit my office in Smithers to complain bitterly about the split shot artists and to demand that something be done to get them off the water. Meanwhile, some of that competitive crowd began spilling over into the Skeena. Still, through all of this, I had yet to see a guide on the Skeena.

The next couple of years on the big water were more typical of longer-term average water conditions. Windows of prime conditions were shorter and started later than one might hope for in order to encounter peak movements of fish. In retrospect, that was a good thing. If 1988 had repeated itself with any regularity I'm convinced it would have been standing room only anywhere near the Kispiox. Still, there were some memorable moments.

It was 1990 when I came to know Bob York (other than by reputation). He was in his prime as a steelhead fly-fishing aficionado by then. The Kispiox was the centre of his universe – after his season on the Dean and before heading south to the Thompson. I had encountered Bob on two previous occasions on Vancouver Island in the early 1980s : once on the Gold River, and again on the Burman. I don't recall we ever exchanged names but I knew who he was. We occasionally saw each other on the Skeena in 1990 while he was searching out the best-looking water below Kispiox in his big green inboard-powered machine aptly named *Rainbow Chaser*. On one of those days, I had anchored the boat out in deep water, far beyond reach by wading. My father-in-law, Harry (RIP), and Brock, age 7 by then, were with me that day. I was coaching both of them rather than doing any serious fishing myself. To put things into perspective I'll say Brock was probably more of a threat to steelhead than Harry was. Bob appeared from far downriver and moved into the tail of the run we were fishing. He promptly dropped anchor and began casting. A dozen or so casts later he pulled anchor, fired up his big inboard, ran up and dropped anchor again, this time within casting distance of us. That was a bit too deliberate for me. We pulled up and headed for the beach, where I deposited Harry and Brock before idling back beside Bob and dropping anchor. We had words – strong words.

Bob did not take kindly to what he referred to as "a two-rod man." With his own history as a plug-pulling guide back in those Jimmy Wright days long forgotten, he felt it was downright evil for me to have gear chuckers on board. But by the end of our talk we had got past all that, and the seeds of friendship had been sown. I remember another day later that season when Bob joined me for a fresh pot of coffee in my boat and was kind enough to share his recipes for fly lines he had spliced together to handle the massive runs on the Skeena. I kept those recipes. For several years after that, Bob would check in with me on his way through Smithers for his season on the Kispiox and again on his way south at the end. He was a monumental source of information on fish, fishing, trends, the guiding scene and much more. Perhaps I never saw the man who gave rise to the stories that were told of him. All I can say is the Bob York I came to know was someone worth listening to. We kept in frequent contact well after he was no longer able to afford season-long sojourns to his beloved Kispiox and even after I moved south from Smithers. Sadly, Bob passed away at age 70 on November 7, 2004. His personal letters and captivating written accounts of almost 30 years on the Kispiox and surrounding waters are among my most prized archival pieces.

Examples of some of the larger steelhead caught in the Skeena River near Kispiox, one by the author's father on September 17, 1989 and another by his son on October 8, 1990.

The 1990s saw the Skeena trend in parallel with every one of its tributaries that held the promise of big wild steelhead. The first pivotal event for the area I liked best was the move by the Kispiox Band to charge a fee to anyone who camped on the flats at the mouth of the river. That included anyone who parked to launch a boat there. The problem was that there was never any structure to that and rarely anyone around to collect any money. Still, you dared not leave a rig unattended if you launched and went fishing without evidence of having met the band's requirements. Ultimately, the opportunity to launch there evaporated, leaving either Glen Vowell, just short of two miles downstream, or Hazelton, a further six miles down. Glen Vowell was a viable option all through the '90s and well into the next decade. If one took the time to consult the local chief prior to launch and request his permission there was never a problem. The only complaint I ever heard from him was about those guides who took big money from their clients to operate on land and water they had no jurisdiction over. None of those dollars ever left the pockets of the guides. The first time around, it took some convincing before the chief accepted that I wasn't a guide. When I confessed to being nervous about leaving my truck and trailer on the beach in full view of the village if the perception would be that I was one of the bad guys, the chief assured me I would be okay. He was true to his word.

The splendid solitude and leisurely pace I once enjoyed on those magnificent big runs on Mother Skeena disappeared too quickly. Before the century was done, jet boats were there on any day that conditions were suitable. Rafts and pontoon boats became common among non-jet owners wanting to ensure that every possible piece of water on both sides of the river was thoroughly searched. It became too much of a disappointment to make the journey from afar only to have to indulge in competition for space and to fish water that never got a rest. The final straw – for me, at least – came in 2010 after an encounter with a local (assistant) guide.

A close friend and I left Smithers for a day trip. We launched at Glen Vowell after stopping by the chief's house to gain his approval. Upriver we went to discover fishermen occupying every likely spot. One guide boat was constantly moving two clients back and forth across the river in what seemed an effort to ensure no one else would get to play before they had exhausted all options. No matter; we bided our time and fished some offbeat water until space was available behind them. The best water had two walk-in anglers working through it. Sometime around noon I could see they had retreated to a perch well back from the river, apparently for lunch. I thought this might be a chance for us to work through that water, so I approached them to see if they would mind. They did not speak English but understood what I was asking and nodded their approval. As I turned toward the river I could see our guide friend pulling his boat into the top corner of the run and depositing his two clients. They were at least 200 yards distant when my partner and I waded into the river and began casting, with me nearest the guide.

Minutes later the guide had covered the gap between us and was screaming expletives at the top of his lungs directly behind me. The gist of it seemed to be that he had this piece of water reserved and we should get our asses out of it immediately, or else! It was the most amazing tantrum I have ever witnessed while fishing. No sentence passed without an f-bomb or three. Threats of bodily harm abounded. I kept casting. Our man finally ran out of steam and returned to his clients. I look back on that day now as the ugly, the bad and the good. The good made up for what we had just endured.

Not long after our diplomatic guide departed, while the two foreigners were still enjoying their break, a fish grabbed my fly with unprecedented violence. Three witnesses to the entire show can testify to every word here. That fish hit as if it had blasted out of the starting blocks for the 100-metre race at the Olympics. There was no pause, no jerk, no surface thrashing when it realized it was hooked, just one tremendous burst headed at 45 degrees downstream to the far side of the river and out. It really was something to behold. To this day I don't know how I managed to get out of the water and chase that fish downriver fast enough to avoid being lined.

When it became obvious this fish was exceptional, we put it on the clock. Well into the fight, the two anglers from above came down for a close-up of what from the foot race and trajectory of the line looked to be evidence of size. The repeated runs of this fish deep into the backing right to the bedrock face on the far side of the river had us all convinced it was huge. The back end of the fly line never got inside the rod tip until almost the end. Never have I seen a fish that came remotely close to taking so many long, powerful runs. Finally I managed to wrestle it from the heaviest water out there against the bedrock bank opposite back to the midstream edge of the large, deep back eddy that separated us. All of us assumed getting it that far would see the fish to hand soon after. It still took more time than I ever imagined to work it to my feet, but the old Sage brownie finally slid it into the shallows. The elapsed time: one hour and 40 minutes! Four of us stood over that fish in utter amazement. My partner summed it up perfectly: "Where's the rest of it?" The fish was 37.25 in. length, 19.0 in. girth. I haven't made a cast into the Skeena since.

The author with "where's the rest of it?," Skeena River near Kispiox, September 14, 2010.

If I needed more than the 2010 experience to strike the Skeena/Kispiox area from my angling destination list, that came in spades in 2013. That was the year a nefarious Kispiox Valley resident and once-upon-a-time guide (at times unlicensed) spearheaded a move to have Kispiox Band members charge access fees for any non-guided angler wishing to fish or to launch or pull out a boat of any description anywhere on the many miles of the lower Kispiox or on the Skeena downstream to at least Glen Vowell. It wasn't just the fees but their differential application that stirred the pot. Guides got by with a nominal annual fee for unlimited use. Band officials I contacted would not disclose what that fee was and claimed not to know if or how the guides applied it to their rates. Local British Columbia residents (defined as living in "the Hazeltons") were required to pay $25 per day or $100 per five-day week for foot access to Band-controlled waters. Non-local BC residents paid $50 per day or $200 per five-day week, and non-BC residents $100 per day or $450 per five-day week. Any raft or other floating device commanded a fee of $100 per day regardless of the residency of the boat owner. I have no idea how residency was determined or what authority Kispiox Band members possess that allows them to prescribe such fees. How such blatant discrimination by self-serving guides got past the Glen Vowell chief I once dealt with remains a mystery. I often wonder who does the accounting following the arrival of new rules such as these. Would one dare ask for disclosure of records? Earlier days when the campsite and launch fees were implemented at the mouth of the Kispiox leap to mind, as does Bob York's piece penned over the winter of 1996–97. It was his well thought out interpretation of the results of the classified waters system that had been implemented in 1990. He referred to it as the guide grab. Good thing he isn't around to witness just how big the grab has become.

The escalating effort and all that accompanied it on those once-favoured reaches of the Skeena around the Kispiox confluence was not confined to that area. Whereas the original intent of the classified waters regulations was to protect substantial sections of the mainstem Skeena from domination by commercial interests, that objective has failed miserably. Today there is no part of the Skeena that isn't travelled regularly by those extracting revenue from its steelhead. Accommodation facilities that are carefully marketed and operated to avoid being prosecuted for illegal guiding just keep springing up throughout the mainstem Skeena. New camps accessible only by aircraft have opened on the uppermost reaches of the Skeena in recent seasons. Helicopters, jet boats and rafts, not all of them associated with guides, leave no reach untouched. The progeny of generations of steelhead that saw virtually no sport fishing pressure once beyond Kispiox now endure a gauntlet that extends to the last little tributary upstream. No one ever pauses to contemplate the potential longer-term consequences of catching and re-catching a steadily increasing proportion of each year's annual return of steelhead.

Even the winter steelhead bound for lower Skeena tributaries are now being marketed aggressively by an array of newcomer guides who seem to think they have discovered the grass beyond the mountains. Commercial activity exploded in 2013, a year with an unprecedented combination of mild spring weather, an ice-free river, no early freshet and a decent population of fish. That perfect storm of fishing conditions reminded me of my 1988 experiences with the summer fish 100 miles upstream. YouTube was alive with clips of guided customers catching big, beautiful winter steelhead on the bars below Terrace. Websites promoting fabulous fishing appeared from nowhere. Though the spring of 2014 heralded a return to normal weather and water conditions, there is no evidence that this slowed the gold rush. None of this water

is regulated such that there is any limit on commercial activity outside the historic summer steelhead run timing window, and that will only make matters
worse for anyone expecting to find the proverbial quality fishing experience
in years to come. All the values that anglers insist are their raison d'être are
being compromised, if not destroyed, in yet another manifestation of the tragedy of the commons.[24]

Opposite page: A buck bug and dry line fish caught by the author at "Obvious," October 4, 1990 following the two Bobs drift trip on the Babine.

12

NORTHERN EXPOSURE

Leaving southwestern British Columbia for points north just as another winter steelhead season was about to unfold on rivers I had come to know since my teenage years took some adjustment. Winter in the northern interior does not afford much steelhead fishing opportunity close at hand. My fishing gear hung on the carefully organized rod rack in the basement of our new abode and collected dust. Office colleagues recognized my withdrawal symptoms early in the new year and promised that as soon as conditions permitted we would make a trip to Kitimat, where the prospects for encountering a fresh winter steelhead were reasonable. The opportunity came in late March.

George Schultze, one of my all-time favourite Smitherites, and I loaded up a couple of small inflatables and a bicycle to deal with a post-drift shuttle and left Smithers early one morning for a day trip to Kitimat, three hours distant. The fishing was less than memorable but it felt good to be on the water and going through the motions. Well along on our drift we were fishing an attractive run when George spotted a moose eyeing us from the streamside willows and alders on the far bank about 100 yards below. Now, I had previously heard about George's hunting prowess. Among his many talents he was a moose caller extraordinaire. I was downstream from him and closest to the moose as he began to talk to it. The moose turned toward us immediately and slowly walked right out into the middle of the river and parked itself facing us, not more than 30 yards away. George continued to make moose talk while I stood there chuckling at both of them. Then I decided we could have even more of a laugh.

I wound up my line that had been hanging in the current below me, reached out and grabbed the leader and pulled the split shot right down against the hook. Then I looked back at George and made some hand signals to the effect of "Watch this." I made a cast over to the far side of the river upstream from the moose and played out line until I could see my float well downstream from where the animal stood. George started to laugh in anticipation of what would follow. Slowly, I retrieved line until my float bumped up against the moose's hind leg. I gingerly drew the float over the leg until I felt the shot-weighted hook make contact. One mighty strike and I was fast to my first moose.

The moose did not react – hardly a surprise given the anatomy of a moose leg. A few minutes of me hauling on the thing while George snuck around behind me to take a couple of pictures was enough for the moose. It trotted out of the river and off into bush, peeling almost all the line off my level wind and wrapping it around enough trees to ensure I was never going to get it back. We still laugh about that encounter. Not bad for an inaugural trip to a new river with a new friend.

Moose on a fishing rod, Kitimat River, late March 1987.

My next excursion to a coastal stream was organized by two other office colleagues, Ron Tetreau and Bill Chudyk, both anxious to convince me there really was good winter steelhead fishing to be found on the north coast. It was April 15, prime time for snooping around coastal streams. What we were hoping for but could never predict from faraway Smithers was to arrive immediately following the onset of the first sustained snowmelt that released the rivers from winter's grip and attracted the early waves of feisty spring steelhead.

Our destination for the day was the Ishkheenickh River, a tributary to the tidal reaches of the lower Nass. I had heard of this river a dozen years earlier from another old friend and ardent steelheader, George Riley, who had been in there via helicopter with the first timber cruising crews in 1968 and '69.

What stuck with me from George's stories was the chinook fishing he had done on some of those early work-related flights. The Ishkheenickh wasn't the only river in the general area around the lower Nass and down toward the mouth of the Skeena that supported a large population of chinook, but it was distinctive in that it produced some behemoths. One memorable tale described an encounter with one of those Ishkheenickh giants in which a handgun was used to dispatch it lest it escape while being wrestled to shore in water too shallow to support its estimated 60 lb. size. Neither the Ishkheenickh nor any of its neighbours today support anywhere near the number and size of chinook that were once prevalent.

On the drive out to Ishkheenickh that April day I learned that neither Ron nor Bill had ever been to the Ishkheenickh either. They had credible informants, though, who had convinced them they had been missing out. Our first stop was the DFO office in New Aiyansh to join up with one of them, a fisheries officer, who had been teed up to serve as our tour guide. We arrived at the end of the old logging road up the Ishkheenickh Valley disappointed to have passed six other anglers along our route. One of them told us a five-person crew from Smithers had left the day before, after spending several days camped on the river. So much for virgin fish.

Spirits dampened somewhat, we wandered upstream past the end of the road access to the first likely-looking run. My first thought was Gold River north. The river looked to be at an ideal flow. There was an ever-so-slight tannic stain to it with just enough turbidity to make fish unwary. It wasn't as big as the Gold River but the colour, the gradient and the substrate in the area we arrived at were remarkably similar. Any disappointment over the traffic we had already seen was quickly forgotten as we fished through that first run. Ron and I released six nickel-bright fish, most of them of double-digit weight. That was all it took to convince me I would spend more time there.

I made two more trips to the Ishkheenickh that year, once in early May and again near the end of the month. The first trip was a family camping excursion with officemate Bill, his wife, Marg, and son Colin. While I wouldn't qualify it as serious fishing, two stories stick out from that visit.

We arrived at the river late on a Friday evening. The best run on the lower river was also the best place to park our campers, literally a cast away from water's edge. Already there was another truck and camper, recognized instantly by Bill as belonging to one of his long-time officemates from Smithers. Naturally, Bill walked over and knocked on the door to say hello. His mate was home but the woman who answered wasn't his spouse. The camper was gone when we arose the following morning. The other story of that weekend is a bit more on topic.

The road layout on the lower river was in the shape of a *Y*. After turning inland from the mainline logging road along the Nass River, the Iskheenickh road proceeded upstream for two or three miles to a fork. The right leg went downstream to a long since abandoned and rapidly decaying bridge two miles downstream. The left leg went upstream and first met the river at the campsite where we were set up. The two access points made for a perfect little drift. I had my Zodiac Youyou with me and worked out a deal where Bill and Marg would ride herd on our three kids while Lori and I did the drift. Before we were even out of sight of our campsite I had landed five chrome-bright fish. On the next corner, barely out of sight, I hooked fish on ten successive casts and landed six of them. Lori's comment: "Is it always like this?" My reply: "No, dear, it isn't."

The final trip to Ishkheenickh in 1987 came on May 22. It involved an overview of the entire drainage by helicopter to assess steelhead distribution and the productive potential of the two largest tributaries. Between that trip and two others, in May of both 1988 and 1989, a benchmark was established that I'm not aware anyone in the steelhead management business has ever attempted to relate to present circumstances.

For perspective, think about two anglers downing tools after bringing 48 steelhead to hand in one day. My familiar line applies once again – I'm thankful I was able to see one more of the province's wonderful steelhead rivers while it (the upper half at least) was pristine and comparatively untouched by angling traffic.

Office colleague Colin Spence on the Ishkheenickh River, May 12, 1989.

By 1990 I was familiar enough with all the prominent winter steelhead streams between Douglas Channel and the lower Nass River to fully understand their nature and put the Ishkheenickh into proper perspective. All supported relatively small populations of late-winter/spring steelhead that arrived in the window between winter and full runoff. In some years the window might be six weeks, in others less than half that. If the winter had been severe and the snow accumulations large, the window of opportunity for access could be even narrower. None of the streams were large and none of them very long. None had all that much fishing water and it was never hard to figure out where the fish were and how strong that year's return was.

These were not productive streams by any definition of the word. Low nutrient levels, high flushing rates, short growing seasons, harsh winter conditions – all of these things conspired against high abundance. What often disguises that reality is being on any of these rivers when you have the water to yourself under good fishing conditions at the right time of year. If you're fortunate enough to experience that, it isn't hard to leave thinking that's the way it always is. Only experience will teach you there is a strong inverse relationship between the number of anglers on the water and catch per angler. The few years of my trips to the Ishkheenickh saw a steady increase in traffic. Road access to the best of it was good in the earliest of those years, so that kept a steady flow of visitors coming as news of good fishing spread. Meanwhile, helicopter-facilitated guiding was becoming a regular event in upstream reaches beyond the point where nature had removed the bridge that once provided vehicle access. There was a pause in the action at one point, though.

When the Nisga'a First Nation was in full-on treaty negotiation mode and taking every reasonable step to pressure governments to deal with their claims, they installed a gate on the logging road along the lower Nass, about ten miles distant from the preferred campsite on the Ishkheenickh. That took care of all but the most innovative anglers and those who could afford an expensive helicopter flight from Terrace. The fish got a reprieve, although there was always some gillnetting of steelhead in the tidal reach of the Ishkeenickh by residents of Greenville, three miles upstream on the other side of the Nass.

The fact that road access to the Ishkheenickh had been put on hold by the Nisga'a only whetted my appetite to find other ways in. I knew from helicopter flights over the area that there was potential for boat access if some way could be found through the maze of logs that choked the river's confluence with the Nass. The added variable there was tides. The only possible way around the

A gillnet strung entirely across one of two channels of the Ishkheenickh River immediately upstream of its entry to the Nass River, April 12, 1991.

logs was a tiny side channel that wound its way around the main clusters of logs on the seaward side. But there was a catch: it didn't carry enough water to accommodate boat passage except on the highest of the high spring tides and only for about an hour on either side of the high slack. Sorting out the timing was relatively easy but there was zero predictability around the river conditions and fish supply that might be found upstream if you did. Still, it was worth a try.

The collection of logs blocking access to the main channel of the Iskkeenickh River at its confluence with the Nass, April 1991.

Gerry Taylor, another of my government colleagues and mentors, and I organized an adventure near the end of April 1991. We arrived at the log dump on the lower Nass, launched my jet sled laden with camping gear and headed off to catch the tide at precisely the right hour. The ten-mile trip down the Nass was easy enough and the small side channel we were looking for was just right for passage around the log jams that prevented any other course. The only problem was one large log right at the top of the side channel just before it joined the main channel of the river above all the logs. That log was a carbon copy of the one I described earlier in the tale of travelling down the Salmon River after seizing the engine in our truck. All we had to do to surmount the log this time was move all our camping gear to the stern, run the boat up on the log until it was teetering, move all the weight forward and slide the boat off the log. The tide was dropping behind us by this point, so our only option was to keep going upstream.

Another couple of hundred yards and we encountered the next problem. A massive spruce tree had fallen from river left across the entire main channel. It was far too large for the 20 in. bar chainsaw on board, so it looked as though we were done. Then we noticed there was a potential way around the impasse if we could clear a path through another small side channel blocked by another series of alders that had fallen into the channel from the eroding bank adjacent. Those trees were small enough for my saw to handle.

It was no small task to cut a swath through that jungle of fallen alders to create a boat-width lane, but we were committed by that point. Upstream travel was nothing compared with the problem we knew was ahead of us coming back down days later, though. On the outside of the bend where the side channel exited the main river, a right-angle turn was forced by yet another cluster of logs. Negotiating the corner around those logs was simple stuff going upstream, but making the same manoeuvre going downstream – to enter a boat-width lane with the river pushing us hard into the logs right at the critical point – promised to be dangerous. We decided we'd deal with that problem when it came. In the meantime, there was fishing to be had.

Once clear of the last of the log problems, getting up the river to the camping area formerly favoured by the drive-in folk was uneventful. It turned out the river was at a perfect height. We lucked out in being there on the downside of the first rain-produced freshet of the spring followed by the first warm spell that sustained snowmelt. Better yet, there was no one but us to check out the fish supply. We had four days of the kind of fishing most could only dream of.

Surmounting the navigation obstacles to gain access to the fishing water on the
Ishkheenickh River, April 27, 1991.

Boat travel on the Ishkheenickh wasn't possible beyond the old washed-out logging bridge about five miles upstream from tidewater and only a third of that distance above our camp. Even at that, the river had to be at a full fishing flow or the rocks would claim any boat that tried. Gerry and I were perfectly happy with that. The water within walking distance of camp was well enough supplied with newly arriving fish; the only real need for the boat was in getting there initially or crossing at camp to fish a couple of runs below.

By day three of our trip the water was dropping slightly and the immigrating steelhead population around the camp waters diminishing. It was time to make the trek upstream to the run I had fished on that very first trip to the river four years earlier. There was still as much as two feet of snow on the shaded old logging road we would be using to get there, but it was well enough packed and crusted on the surface to make travel manageable in the cooler early-morning hours. The promise of fish anaesthetized us to the prospect of what the return trip would be like after the midday sun had softened the snow.

It was downright hot by the time we arrived at our destination. We were already breaking through the snow on every second step by then. I'll not forget shedding waders and all but wringing sweat out of my clothes when we finally made our way out onto the snow-free flood plain. The fishing was well worth the effort. I'm quite certain Gerry caught more fish over the next few hours than he ever had before or since in a single run. I took a lot of pictures.

Just about the time the catching finally slowed down and we began to think about the death march back to camp, a strange sound could be heard far downstream. The only thing we could come up with for a match was a skidder but that didn't seem likely given the gated road and the ongoing Nisga'a treaty issues. The sound persisted for ten minutes or so before a red object came into view several hundred yards downstream. We knew it couldn't be a boat but we could see plainly the noise maker coming up the river channel right toward us. Two or three minutes later the red thing put ashore and shut down, and the operator and his dog, both wearing ear protectors, stepped onto the beach. I recognized the operator as Cress Farrow, the man who ran the building maintenance services at my office in Smithers. His mode of transportation was a hovercraft! He and his party had taken the same route we did to get into the river and discovered my boat and our campsite right where they were headed. They then unloaded all their gear and Cress set out to find out who was on the river and where. The rest of his party were in a jet-equipped Metzler inflatable that remained around the camp waters while Cress was on the hunt.

The hovercraft belonging to Smithers resident Cress Farrow shortly after it arrived at the run being fished by Gerry Taylor and the author on April 29, 1991.

Dumbstruck hardly does justice to our impression of the hovercraft. Its ability to get to where we were was nothing short of amazing. Cress was totally nonchalant while explaining its operation and utility. All I could think about while he and Gerry chatted was the death march ahead of us. When Cress put the ear protectors back on his dog and was about to head back downstream I couldn't resist: "Say, how many bodies will that thing carry?" Cress replied, "Oh, do you want a ride?" Minutes later, three of us and a dog were heading downriver on the ride of a lifetime. I couldn't believe that machine could just go right over the top of huge boulders that were barely even wet and cut corners over dry bars on the inside of river bends. Fifteen or 20 minutes via hovercraft or three hours of ploughing through soft snow in neoprene chest waders? Not much of a contest there.

Gerry and I didn't fully disclose our fishing results when conversing with Cress about the days prior to his arrival. We had probably educated almost every fish between our camp and where he found us, but we spared him the "ya should have been here yesterday" stories. After all, he had saved us from ourselves, not just with the hovercraft ride but also with a mini-logging operation.

When we were first talking to him upriver, we told our tale of getting past all the logs and expressed concern over the trip out. Cress told us we wouldn't have to worry. His party had taken care of the problem. I learned later, long

Gerry Taylor with one of many caught shortly before the hovercraft arrived,
Ishkheenickh River April 29, 1991.

after returning home from that trip, that Cress and his friends were the same
crew that had been on the river just before my very first visit four years earlier.
They had been making annual trips to the Ishkheenickh for years and were very
familiar with the problem of getting past the tidal reaches and safely beyond
the braided, log-filled channels immediately upstream. They had logger-size
chainsaws and pee-vees that made short work of the big spruce tree and others
that had blocked our way and forced us into the side channel route coming in.
Thanks to them, our trip out was a cakewalk.

Boat access improvements courtesy of the Cress Farrow party that arrived at the lower
Ishkheenickh River several days after our access adventures in the same area.

I made one other observation on the April 1991 trip. It speaks to the volatile nature of the north coast streams. Sometime during the early spring that year there was a large accumulation of ice farther up the Ishkeenickh Valley than our fishing adventures took us. It broke loose before Gerry and I arrived, but the aftermath was right there in front of us. Fellow staffer Colin Spence and I had seen evidence of another event when visiting the river at the end of April 1989, but the signs were nowhere near as obvious that year. In 1991 there was a wall of ice and snow, with much embedded gravel and cobble substrate, deposited at the end of the straightaway section of river just downstream from our camp. It had come roaring down the river and spread all across the floodplain at the point where the river turned sharply left. The accumulated material was ten feet above water level on both sides of the river. In 1989 Colin and I found sand, gravel and large accumulations of woody debris similarly deposited on the lower river floodplain to an estimated elevation of eight feet above the water's surface. There was also a layer of fragmented ice deposited in the bush well back from river's edge. The consequences of those sorts of events for rearing juvenile steelhead must have been catastrophic. Two occurrences over six seasons is an indication of just how limiting an environment these rivers present for a species with an extended freshwater rearing component in its life history.

My last Ishkheenickh adventure came in 1992. The gate on the lower Nass road was still a fact of life, so the options for getting to the river were unchanged from a year earlier. Our boating escapades of 1991 were not something I cared to repeat, even if my opportunity to make the journey had coincided with the tides. But it didn't, so that left the gated road as the only game in town. I had a plan, though. Take the truck and camper to the log dump site on the Nass, overnight there and get an early start via mountain bike to the old campsite run; from there, drift the river in my Zodiac down to the takeout at the end of the other spur, the same drift I had made with Lori five years earlier. I had Brock with me, so that made the mountain bike trip a little more onerous. Two bodies, fishing rods and an inflatable on one bike made for an interesting trip. One way was about ten miles. Thankfully the terrain was level all the way.

We soon discovered what two years of zero maintenance can do to roads in a coastal environment. Whereas the mainline down the Nass beyond the gate was relatively unchanged, the spur that turned up the Ishkheenickh had been reclaimed by the ever-industrious beaver. A couple of strategic culverts were dammed, leaving water flowing over the road and eroding several areas.

Evidence of major ice sluice outs on the lower Ishkheenickh River in April 1989 and again in April 1991.

Much water-borne debris littered the road. We needed the waders we were wearing. In my diary I referred to the day as "operation access."

It took about an hour and a half to make it to the river from the Nass turnoff. It was lower than ideal, but the campsite run where we started yielded four fish nonetheless. Much to our surprise two other parties on trikes arrived while we were there but they went upstream when they saw us. There was also a helicopter ferrying a couple of rods up and down the river. Drifting wasn't an exciting prospect at that point but we inflated the Zodiac and made off downriver anyway. That proved to be a bad decision. Three more fish along the way and no other anglers, which was reasonable, but the river was infested with logs that put the original takeout destination out of reach. Instead, we pulled out well upstream, readied the gear for travel and went overland to intercept that same spur Lori and I had used in 1997. My diary says "nightmare." Nature had reclaimed that road as well. It took us three hours to get from water's edge to the truck. So ended day one of our weekend.

Fiends for punishment, we repeated the bike-in show the next day, raft included, but only to fish the far side of the campsite run and a piece below it. Our tally was two fish, both old and stale. If the number of fish landed was the metric, those two days represented an all-time low for me. It was a good reminder of the fact that fish and fishing are never constant. We weren't the only ones to be humbled. As we were preparing to leave the campsite run and head back to the truck, three other mountain bikers arrived from upriver. They had come in behind us and failed to notice we were on the other side of the river just downstream. Having fished all the best runs upstream and drawn a blank, they figured they still had their ace in the hole: the campsite. They were disappointed enough to discover we had already fished the run. The news that it wasn't exactly full only made it worse.

The author's son with a typical winter steelhead he caught in the lower Ishkheenickh River on April 25, 1992.

The author's son exhibiting the mountain bike used to gain access to the lower Ishkheenickh River, April 26, 1992.

13

STEELHEAD PARADISE REVISITED

In late November 2013, I received a request from the editor of a prominent fly fishing magazine. He had read my Skeena steelhead book and was particularly interested in my remarks about the early days on the Sustut. He asked if I would consider writing a short essay for his winter issue about the time I had spent in Sustut/Johanson country in 1972. Because the press deadline was imminent, the request meant I would have to readjust my schedule somewhat. There was loose talk of compensation of about $300, which I understood was normal for such things. I agreed and sent a three-pager his way a couple of days later. A response thanking me for the article came back quickly.

When the winter issue hit the newsstands I went looking to see if my piece was included. It wasn't, nor did it appear in the magazine's spring edition. At that point I went back to my friend the editor. He tripped all over himself apologizing for not having got back to me and then went on to explain, in so many words, that he didn't want to publish what I had sent him because it wasn't a glowing account of fish, fishing and pristine loveliness that would appeal to his subscribers. We left off with his offer to pay me $50 and hold the piece for possible future publication in a truncated version. I told him he could do whatever he liked with the article and keep his money. Compromising truth for the grandiose sum of $50 is hardly something I would be proud of.

The Sustut and Johanson I visited in 1972 and described for my editor friend were not what I had anticipated based on John Fennelly's *Steelhead Paradise*. On arrival in late September that year I was immediately struck with how small the two rivers were. I suppose that if I had applied an analytical eye to Fennelly's photos and accounts I wouldn't have been quite so surprised. I surmised that his visits must have occurred during much higher flow stages, but that wasn't much comfort to me at the time. Every fish present in the fabled junction pool was clearly visible. It wasn't paved with them, either. As I walked both the Sustut and Johanson above their confluence in the days following, I could easily cross either river virtually anywhere. There were only scattered fish in those areas as well. Surely they were lined up by the dozens or hundreds in the quiet water upstream, in the lower bays of the two lakes so beautifully described by Fennelly.

As fate would have it I never got to see the lakes whose outlet reaches were Fennelly's favourite playground. The entire area fell victim to early winter-like weather. Air temperatures plummeted and ice was beginning to pile up against the counting fence at camp. That's all it took for the man in charge to pull the plug. On the morning of my fourth day on-site the helicopter returned to whisk us away before the weather could take us hostage.

That inaugural Sustut adventure was hardly a disappointment, though. Hiking along two sparkling clean rivers alternately tumbling and meandering through subalpine uplands painted in fall colours and laced with the trails of moose and caribou, stumbling over a grizzly's cached caribou kill, witnessing moose in the peak of rut, midday skies unimaginably blue, northern lights, periods of intense silence interrupted by howling wolves, fish as wild as they come almost a mile uphill and 400 miles from salt – all of these made for an experience never to be forgotten. My only lingering regret is that I didn't go equipped with the camera gear I should have to record an ecosystem while still at its best.

In the foreword to his 1963 *Steelhead Paradise*, Fennelly wrote:

Although it is still predominantly an area of rugged mountains and trackless forests, each passing year witnesses a gradual encroachment of civilization upon the wilderness sections of this country. Within another decade or two most of its primeval charm is certain to disappear.

How foreboding. Less than a decade after Fennelly's last visit to steelhead paradise, a railway was under construction right beside the lower Sustut's major tributary, the Bear River. Months later a bridge crossed the Sustut at the confluence. At roughly the same time a mining development road penetrated the upper drainage from the southeast. A major mine commenced operation not long afterward. The pristine loveliness of Sustut and Johanson was lost forever. Today one can drive to the Junction Pool.

The entire Sustut system upstream from the Bear confluence has been closed to fishing since 1973, when the influx of badly behaving railway construction crews demanded that measure. This constrains the non-Aboriginals among us, but First Nations claim it as their territory and with that goes rights to harvest all things with fin and fur. They love the road. A counting fence not far below the Sustut/Johanson confluence has been a fundamental feature of the stock assessment scenario for Skeena steelhead since the early 1990s. That too serves some First Nations folk well. Steelhead and salmon are much easier targets when kegged up on the downstream side of a weir or when they are sitting ducks in a live trap. The steelhead count through that fence averaged slightly better than 700 fish, with an upper limit of about 1,500.

The railway that was supposed to "open up the North" in the 1970s was abandoned as a bad investment less than a decade later, but by the mid-1990s the sawmills in distant Prince George were out of logs. The government solution: restore the abandoned rail line. A bad investment quickly became a good one when forest industry jobs were at risk and votes needed to be bought. With the logging came the roads and bridges throughout the lower Sustut and out to the Skeena. There is still no road access into the lower drainage, where two steelhead lodges now squire anglers around the Sustut below the Bear as well as up and down the Skeena beyond the Sustut confluence. However, the road networks and bridges developed once the railroad was upgraded and crews and equipment were imported have scarred the landscape just the same. The quest for minerals, oil and gas continues to erode what little remains of virgin territory in the upper Skeena drainage.

If the industrialists weren't already enough, jet boat jockeys arrived in 2013 to enjoy the Sustut-area wilderness experience by being the first to roar from Hazelton through the canyons and cascades of the Skeena to well beyond the Sustut before returning to run the Sustut itself past the Bear confluence. They left after a day or two at a steelhead lodge to conquer the Babine on their way back to Hazelton. All of this was proudly advertised in a feature article in a prominent Idaho newspaper. Just what the Sustut doctor ordered!

It took less than half a human lifetime to transform a piece of virgin wilderness so as to accommodate everything but the fish and wildlife dependent on it. Fennelly saw it at its best, and I at the edge of the downward slope he predicted. Fifty years ago I was entranced by the vision he created. Today I appreciate him even more for his seminal book, without which much of the history of the Sustut and Skeena would have passed unnoticed.

Having tracked the entire evolution of the Sustut steelhead fishery through personal visits and association with the people and organizations involved, I see the river differently than do the guides who have rotated through there and the anglers who may have fished it for a week. It is not nor ever was close to the Babine in terms of the number of steelhead it has supported at any time during the period of record. I can find no evidence it was ever any different, even before Fennelly. The fish he revered in the uppermost reaches of the system

are now understood to move through the lower river before the fishing activity
concentrated on those reaches by guides begins. The upper river fish were
never large either. That ought to be crystal clear from Fennelly's accounts, but
if it isn't, the monitoring of the fish passing the weir near the Sustut/Johanson
confluence over the past 22 years removes all doubt.

In 1986, when the first comprehensive investigations of Sustut steelhead
were undertaken, a reasonable understanding of the river's fish emerged. It
is worth remembering that 1986 was the third of three successive years of
an unexpectedly high abundance of steelhead in all the major Skeena tributaries. Obtaining sample sizes to derive valid results was never a problem.
More sophisticated sampling and monitoring has been done since but has
served only to dot the i's and cross the t's on the original work. The best of
the fishing was always in the first few runs below the Bear River confluence,
which hadn't changed from 1972 except that the dollies that had been there in
nuisance abundance at the same time 14 years earlier were gone. So too were
the grizzlies that had once patrolled the Bear, fattening up on the formerly
abundant chinook. It had been established years before that the lion's share of
the fish found in the lower Sustut in September and October spawn in the lake-
headed Bear system, so it was no surprise to the 1986 crews that the largest
concentrations of fish would be found in the confluence area.[25] Those fish were
significantly larger than their upper Sustut and Johanson counterparts, too,
although not Kispiox large. Every fish bound for Bear River obviously passes
through the lower Sustut but, like their upriver-bound counterparts, they didn't
linger in the lower river.

Steelhead lined up at the outlet of Johanson Lake on September 30, 1986.

My only opportunity to see the still relatively unspoiled lake outlet area at Johanson came on the final day of that September 1986 trip when I joined the local staff for their weekly flight into the area. Just as Fennelly had described a quarter-century before, steelhead were lined up in regimental fashion across the shallows just above where one would draw a line between lake and river. There weren't hundreds, as Fennelly implied was common during his years, but there were at least two or three dozen. It was comforting to know that bit of the ecosystem was still at least partially intact.

The one feature of the Sustut fishery that puzzled all of us involved in the early investigations was the location of the two steelhead lodges. The lower-most lodge, Steelhead Valhalla, is about four miles upstream from the Skeena but 16 miles downstream from the Bear confluence. Suskeena Lodge is five miles upstream from Valhalla and 11 miles below the Bear. The original mode of access to Valhalla was mostly via floatplane to the river directly in front of the lodge. That being as risky as it was, the preferred method of getting to both lodges became fixed-wing aircraft to an old airstrip once used to supply railway construction crews working the stretch along the Skeena just upstream from the Sustut confluence. Of course, helicopters were an even better choice but not always affordable. Home-built speeders rotated clients between the airstrip and the lodges. That may have simplified getting to the Sustut from the outside world but it didn't begin to offset the problem of daily transport of sports from lodges to the areas where the fish were most likely to be found.

At low water, the Sustut reaches near Valhalla are low gradient, braided, shallow and less than ideal for either boat travel or fishing. Conditions improve with increasing distance upstream but even Suskeena is not strategically located for the best access to fish and fishing. Anyone who has ever navigated the Sustut from the Bear confluence downstream late in the day when the sun is in their face knows of the hazards of boat travel during low water periods. In my experience it was never hard to find rocks well marked by aluminum that had once belonged to jet shoes. I know from conversations with the original proprietors of Suskeena Lodge just how much of a problem it was for them to get their clients to where the best fishing was day after day. They rarely went downstream. The abandonment of outboard jets by the Valhalla proprietors in favour of inboards installed in longer, lighter, shallower draft custom-made sleds was a first among Skeena-area steelhead guides. They did that for obvious reasons.

Top: The author with the largest of the many steelhead caught during the steelhead population estimation exercise on the Sustut River in September 1986.
Bottom: Typical female steelhead in the upper Sustut River, September 30, 1986.

Opposite page.
Top: Fish tagging crew in action on the upper Sustut River, September 30, 1986.
Bottom: Ron Tetreau fishing the Sustut River near the Bear River confluence, September 28, 1986.

Of all the Skeena tributaries I have been fortunate enough to spend time on, none matched the Sustut for scenic splendour when the valley was still relatively intact. Visit it today and bear witness to progress. Alternately, choose the contemporary technological route and compare the photo below with the valley's appearance today on Google Earth. Roads, airstrips, power lines, cut blocks, abandoned camps and drilling platforms and massive mining developments are all there out of sight and out of mind of the masses of people who live off the avails of it all. The assault on the lower Sustut is troubling enough, but the transformation of the ultrasensitive, fragile environment of the headwater lakes area and the "Junction Pool" of Fennelly's day is even worse. Wounded landscapes at such latitude and elevation will never heal. Add one more of British Columbia's once-upon-a-time treasures to my list of places I'm thankful to have seen when the opportunity was still there.

The lower Sustut Valley as it appeared halfway between the Bear confluence and the Skeena in September 1991.

14

REALLY BIG FISH

One of the benefits of being in the centre of the big steelhead universe for so many years was that I had the opportunity to chase down rumours that surfaced within the angling community. Whenever a worthy candidate reached me, I did what I could to verify it by searching out either the person directly involved or an eyewitness. Most of those efforts were successful, and a record was made of exceptional fish. Some of those are shared here but not before a few more accounts of celebrated fish from yesteryear. I'll preface this by acknowledging there is little doubt there have been fish of similar or, possibly, greater size caught but not given their due in any official context. Some were killed, others not. In recent years, of course, we don't kill any wild steelhead so we're left with estimates rather than autopsies. This injects an element of uncertainty if the objective is absolute accuracy, but thankfully, that is no longer important. I should also say there are many accounts and photographs of large, angler-caught steelhead available at the click of a mouse. Anyone can look those up and consider them for what they are worth. I'll make no effort to either acknowledge or verify the accuracy of the overwhelming majority of them. The fact that many of the entries claim to be International Game Fish Association records but don't appear anywhere in the output of that long-acknowledged clearing house for all such assertions says much about the veracity and reliability of the Internet.

I think it is safe to say the historical records of angler-caught steelhead leave little doubt that the largest specimens on the planet belonged to the Skeena system. Yes, the occasional large fish has been taken from the Thompson River, Idaho's Clearwater and some of Washington's storied rivers, but not with anywhere near the frequency that has occurred in Skeena country. Long before the mainstem Skeena fishery developed, it was the tributaries that attracted all the attention. The Kispiox was foremost among them because it was the first of the major tributaries to become readily accessible by road. The fact that the Kispiox was also the river whose fish took top honours several times in the 1950s in the annual fishing contest sponsored by *Field & Stream* magazine, an enormously popular information source at the time, cemented a reputation that has never been refuted in the 60 years since. The only question is whether some of the monster fish taken from the Skeena River bars below Terrace or from other areas of the Skeena farther upstream were bound for the Kispiox. Today's sophisticated DNA analysis could answer the question if ever necessary.

The angling world is well aware of some of the leviathan steelhead already on record. The International Game Fish Association lists the world record steelhead as 42 lb. 2 oz., caught by 8-year-old David White of Seattle on June 22, 1970. He was fishing out of a rubber boat in Bailey Bay near Bell Island, Alaska. There are several interesting features about that catch. First, it was in the ocean. It can't be overemphasized how rare an event it is to encounter any steelhead at sea, much less a world record. Second, it was nowhere near any river known to produce large steelhead. Southeast Alaska is the last place one would ever go to find a fish of such proportions (especially at that time of year) so it must have been bound for somewhere in northern BC. The Skeena system seems the most likely destination, and the timing is about right, but Bell Island is far removed from any reasonably direct route from the central north Pacific steelhead-rearing grounds to the Skeena. The BC river closest to Bell Island is the Unuk, which has never been known to produce any number of steelhead. Third, the fish was assumed to be a chinook salmon until the taxidermist who was mounting it became suspicious and consulted University of Washington experts, who correctly identified it as a steelhead. The only dimension available for that fish was its length: 43 in.. The picture of the mounted specimen does not do the fish justice if that length measurement is correct. A 43 in. steelhead would have to have a much larger girth than that displayed by the mounted specimen to top 42 lb. It is possible the length measurement was taken after the fish was mounted, in which case the shrinkage inherent in the skin mounting process could explain the discrepancy.

The other notable steelhead commonly referenced as world records at one time or another include Chuck Ewart's 36 lb. Kispiox fish, taken on October 5, 1954, and Carl Mauser's fly-caught record fish of 33 lb., also from the Kispiox, on October 8, 1962. An interesting sidebar to the Ewart catch is that it was the second time that year a new world record steelhead had been weighed. The first fish came in at 31.5 lb. It was caught by Arthur Mowat of Hazelton on October 1. Mowat's friend Jeff Wilson, also of Hazleton, had held the record previously with a 30.5 lb. fish caught on September 6, 1953. Mowat's fish created such a stir within the sport fishing community that the *Vancouver Sun*'s Lee Straight dropped everything and headed for Kispiox to verify the rumours. According to his columns of October 4 and 5, his whirlwind trip involved tracking down both Mowat and Wilson, fishing with them for a day and returning to Vancouver, all in the space of 48 hours. Mowat gave his fish to Straight, who subsequently had it mounted for display in his study. Two days after Straight had left Kispiox, Chuck Ewart landed the next world record.

Ewart's account of his fish appeared in the Prince George newspaper at the time. He reported that it took him an hour and 45 minutes to land it on spinning tackle. Its length and girth measurements were recorded as 44 and 24 inches respectively. The mount of that fish is still on display in the Northern Hardware store in Prince George.

The *Field & Stream* records of the largest steelhead caught and registered in 1954 include both the Mowat and the Ewart fish. However, Mowat's fish took third place, not second as expected. That was due to a 31 lb. 14 oz. steelhead caught by Wendell Henderson in the Babine River on October 24.

Mauser's personal account of his fly-caught record was intriguing. A copy of his diary entry that I have had filed away for many years tells the story:

Went to claybank about 1 PM. Fished through with 2 patterns but nothing happened so drove to Bed Rock. I had fished thru about half when I lost a fly so retied with a big special and after 2 or 3 flips caught and landed a new world record fly steelhead – 33 lbs – 42 ½ inches long – 24" girth – ridiculously easy – less than 10 minutes – will have it mounted by someone.

One of the more amusing stories of large Kispiox fish references an Oblate missionary from Hazelton who caught what would have been a world record steelhead at the time (October 1952) except that it was never officially weighed. The story that appeared in the *Prince George Citizen* months afterward held

Carl Mauser's 33 lb. fly-caught steelhead, Kispiox River, October 8, 1962.

that the missionary took the carcass to a friend who was going to can it for him. The friend recognized immediately what the missionary had caught and put the headless, eviscerated specimen on a scale. It weighed more than 29 lb. An official of the BC Game Commission was contacted and estimated the round weight of the fish would have been about 35 lb.

The accounts of large steelhead caught from the heartland of it all, the Kispiox, came fast and furious in the mid-1950s but tapered off markedly over the next 30 years. Occasional stories of exceptional fish from Skeena country emerged during the 1960s and '70s but, with the exception of 1966, the frequency of encounters did not match that of the mid-1950s. The *Field & Stream* records for fly-caught steelhead that year revealed the top three prizes were taken by Kispiox fish weighing between 26.5 and 30.2 lb. In addition, one remarkable fish was caught in a commercial fishing gillnet. That one was shrouded in uncertainty but still deserves attention. Once again, a Lee Straight column was involved.

According to Straight, a 43 lb. steelhead was caught by commercial fisherman Kichizo Hana near Port Simpson (roughly equidistant from the Skeena and Nass rivers) on July 25, 1966. The fish was delivered to one of the fish processing plants in Prince Rupert, where it was recognized immediately as a unique specimen and reported to fisheries authorities. Arrangements were made to ship the frozen carcass to the province's chief of fisheries, Ed Vernon, in Victoria shortly afterward. Vernon and his second-in-command, Ron Thomas, reported that the fish weighed 43 lb. when received but was frozen and glazed; thus, it likely actually weighed slightly less. It was 46.75 in. long with a girth of 25.75 in. Vernon and Thomas made arrangements with the Royal British Columbia Museum in Victoria to have the fish mounted for display in that world-class facility.

In 1972, early in my career in the provincial fisheries management business, Ron Thomas told me about this huge steelhead and took me to the museum to view it. As best I recall, the story from museum staff was that the model over which the skin of the fish was to be secured had been damaged, rendering the product unsuitable for display. Curious about the ultimate fate of that fish, I contacted museum authorities early in 2016. Their response was they had carefully searched all their records and stored specimens but were unable to detect any evidence they had ever had such a fish in their possession.

Two other behemoth steelhead should be mentioned here. One was a 35 lb. fish taken in the Kispiox River in October 1969 by Smithers resident Wally Booth; the other was an alleged 37.5 lb. fish taken in the Babine River by Burns Lake resident Tod Lowley on September 12, 1976. Both of these fish were referenced by Lee Straight. The latter does not get my vote for legitimacy. First, it never surfaced in any other reports, despite the fact that it would have been the freshwater-caught record if, as stated by Straight, it was weighed on a certified commercial scale and witnessed by a conservation officer of the provincial government. Second, the length and girth measurements given at the time of weighing were 39.5 in. and 20.5 in. respectively. Those dimensions would describe a fish of not more than the low 20s range, obviously nowhere near the stated weight. Enough said.

A second flurry of catches of exceptionally large steelhead occurred in the mid-1980s. Looking back from the perspective of one deeply involved in the professional side of steelhead management, I'll offer an observation or two.

The mid-1950s matches the timing of the demise of the Babine Lake sock-eye following the catastrophic slide that occurred on the lower Babine River in 1951. The commercial fishery at the mouth of the Skeena River was curtailed significantly in years following the slide in order to protect what remained of the Babine sockeye, the commercial fishery mainstay, and to assist in the stock rebuilding efforts. Ultimately the solution was the Babine spawning channels that became a nail in the coffin of many other stocks and species that min-gled with sockeye on their return to the Skeena. That is a well-known story that need not be repeated here. The point is the prevalence of big steelhead through the mid-1950s looks to have been partially related to less harvest pres-sure by the commercial gillnet fleet. The other potential contributor was opti-mal ocean-rearing conditions, which manifested in both survival and growth. Brood years contributing to high adult abundance are generally also years that support large-size-at-return adults. The ocean environment of steelhead was a complete unknown in the mid-1950s. That didn't change markedly over the next 30 years, but we do know the coast-wide abundance of steelhead in the

mid-1980s was at levels not seen in decades. The explanation at the time was
unprecedented survival at sea. Sure enough, the high survival was accompanied by excellent growth, and the next pulse of large steelhead showed up in
angler catches. The commercial fishery harvest rate of those fish was undoubtedly high (40 to 50 per cent is not an unrealistic estimate) but the abundance
was high enough in those years to still see good fishing and big fish on the
upstream side of the nets. One can only dream about what may have been in
the absence of the nets.

Among the exceptional fish that came to my attention in the mid-1980s were a
38.5 lb. specimen caught by Andy Bouchard of McKenzie at Esker Bar on the lower
Skeena on September 13, 1986, and a 42 lb. fish caught by an unnamed woman
from Alberta at Shames Bar on the lower Skeena on Labour Day weekend in 1984.
Both fish were weighed on site with hand-held spring scales shortly after they were
caught. The larger was reported to have also been weighed at the Terrace Co-op,
where the figure of 42 lb. was confirmed. The Bouchard group reported catching a
second fish the same day that was only slightly smaller than the first.

Two other honourable mentions from the mid-1980s also came from the
Kispiox. One was a widely reported fish estimated to weigh 37 lb., caught and
released on October 1, 1985, by Clay Carter of Ketchum, Idaho. The other was
an alleged 40 lb.-plus steelhead caught by Fay Davis on September 11, 1984.
Davis's fish measured 45.5 in. long by 26 in. girth. Interestingly, the report on
this fish did not hit the airwaves until a drama-filled article written by Davis
himself showed up in an Oregon-based sport fishing magazine nine years later.

Bob York penned a 17-page piece titled "Big Northern Steelhead" in 1996,
not long after his last season in Skeena country. In that engrossing account of his
long experience on the Kispiox, he spoke of a number of encounters, personal
and otherwise, with steelhead of legendary proportions. One point made by York
was that the catch-and-release era had given rise to an increase in the frequency
of stories of exceptionally large steelhead. York's personal tally of steelhead
weighing 30 lb. and more was eight, the largest of which he caught and killed on
September 17, 1973 – and not on fly tackle. It weighed 34 lb. and, as of 1996 at least,
the mount was still on display at the Avid Angler Fly Shop in Seattle.

York's contention about the catch-and-release era giving rise to ever more
frequent reports of massive steelhead didn't go unnoticed by others, myself
included. I tended to be skeptical of such stories unless they were corroborated
by at least a good picture. That rarely happened. There was one encounter I
did pay attention to, though. Daniel Daigle, who worked as a guide for Skeena

River Fishing Lodge, sent me a photograph of himself cradling a fish a client had caught near Esker Bar on September 17, 1993. By that time, all wild steelhead retention was forbidden in the Skeena. The only data accompanying the photo was the fish's length: 48 in. Clearly a fish of that length had to be at least in the upper 30 lb. range. Based on careful analysis of the photograph and some proportional calculations I was able to perform, the fish's estimated length was exactly as reported by Daigle. It is worth noting that 1993 was the third successive year of desperately low returns of steelhead to the Skeena system. Under those circumstances it is reasonable to assume that the survival rate and growth of fish expected to return in 1993 were poor, thus making the odds of catching a fish as large as the one held by Daigle exceedingly low. I would also suggest that the girth of Daigle's fish was less than what was normally observed on the Skeena's heavyweight fish. This may have been an indication that the fish was running low on groceries during the latter stages of its ocean residence. If that fish's condition factor had been more typical of the largest Skeena fish, I'm certain it would have exceeded 40 lb.

The incidence of truly large steelhead dropped off with the overall decline in steelhead abundance throughout the Skeena system in the 1990s. But, just when the world was becoming convinced that the best of whatever swam those waters was already in the history books, 1998 arrived. In my Babine and Skeena chapter I described that year as one for the ages. I'll expand on that characterization with a bit more evidence here. It's important to remember there was no commercial fishery at the mouth of the Skeena that year, so the entire steelhead population approaching the river received a free pass to the sport fishing territory upstream. That circumstance was not unlike those of the 1950s, when Kispiox record breakers made headlines.

On August 1, 1998, the test fishery that operates under contract to the DFO to estimate the abundance of salmon and steelhead escaping from the commercial fishery immediately downstream called in to the Prince Rupert staff responsible for the contract.[26] They reported an extraordinary occurrence worthy of pictures. Three monster steelhead were picked from their net that day. Two were kept, the other tagged and released. The morts were weighed on a spring balance identical to the one I used, so I have complete confidence in the weights recorded at the dock shortly after capture. One was 42 lb., the other 38 lb. The released fish was reported as a twin of the smaller of these two. Fisheries biologist Steve Cox-Rogers of DFO, Prince Rupert, himself an avid steelhead angler, examined the fish on August 4 and noted their

Three exceptionally large steelhead caught at various times on lower Skeena River bars between 1984 and 1993.

dimensions as 43.5 in. by 26.5 in. for the heavier of the two and 44.5 in. by 26 in. for the other. Steve's message to us in Smithers was that these numbers may have been slightly reduced through shrinkage and fluid loss in the fish over the time between their capture and his examination.

Something to keep in mind here: the two fishermen hoisting the fish had more than passing experience with gillnetting in the Skeena. The Kristmanson family's association spanned two generations, while Robert Johnson was the primary test fishery operator for at least 15 years prior to 1998. Just as the 43 lb. (?) steelhead caught in a gillnet in 1966 had attracted immediate interest, so too did the three fish taken by the test net 32 years later. For these veteran gillnet fishermen to report their August 1 catch such as they did stands as a testament to just how rare an event that was. When one considers there has been 140 years of commercial fishermen blocking the entrance to the Skeena with successive waves of gillnets on more days than not when steelhead are arriving, it is not hard to imagine that many large steelhead have been caught. That aside, I'm of the opinion that all fishermen like to talk about big fish and I doubt any steelhead as large or larger than those already on record have escaped notice.

The two extremely large steelhead caught by the Department of Fisheries and Oceans' contracted test fishing vessel at the mouth of the Skeena River on August 1, 1998.

The best is saved for last. When still living in Smithers I came to know a real estate agent, Donna Jeffries, who had an appreciation for steelhead fishing. One day she dropped by my office with a picture she knew I would be interested in. It was a photocopy of an old Polaroid picture she had noticed while at the home of some Kispiox Valley clients. She had borrowed the Polaroid long enough to make a copy, which she kindly left with me. The fish pictured was unlike anything I had ever seen. I was impressed, to say the least, but I didn't do more at the time than file it away with the rest of my collection of big fish photos.

A dozen years later, while sorting through photographs suitable to publish in my first book, I revisited that photocopy and showed it to my son. He noticed immediately the name scrawled at the bottom, which belonged to the mother of a classmate and close friend of his when we'd lived in Smithers. That kindled my interest in learning more about that special fish. Enter Facebook. It didn't take long for Brock to locate his friend, who went back to his parents for the story. From India in March 2011 came a detailed transcription of the conversation between my son's friend and his father, Fred Pitzman–the angler as well as the photographer.

The fish was caught in the Walker Run on the morning of October 11, 1976. Fred was using a stout 9 ft. Berkley rod equipped with a DAM Quick spinning reel, which was loaded with 17 lb. test line to which a large Krocodile lure was attached. Fred recalled that it took somewhere between 30 and 60 minutes to land the beast, after which he lugged it home, took a picture of his wife, Pam, holding it and then cut it into steaks that he put in the freezer. The Pitzmans' neighbours were well-known Kispiox Valley folk, Joy and Gene Allen, who had been building a house that winter. When the Pitzmans left to travel in far-off lands shortly after, they offered their house to the Allens as interim accommodation while the latter family continued to build. Fred told Gene to help himself to whatever was in the freezer. The story

Pam Pitzman holding what the author believes is the largest ever angler-caught steelhead. The fish was caught by her husband, Fred, in the Kispiox River on October 11, 1976.

goes that sometime over that winter the Allens ate some very large salmon steaks. Someone may come out of the woodwork with evidence to the contrary, but until then, I'm of the opinion that Fred Pitzman caught a steelhead in 1976 that would have been the hands-down winner of the title Largest Ever.

Before leaving the subject of big fish I'll say a thing or two about the business of estimating weights based on length and girth measurements. In my view, Bob York's skepticism about the veracity of new age anglers' reports of steelhead that allegedly eclipsed in size the largest fish killed and weighed in the days when that was entirely acceptable was well founded. Those latter-day claims commonly stemmed from fervent beliefs about what a 40 in. steelhead weighs and/or the estimates of weight derived from formulae now broadly applied. Over a professional lifetime of handling thousands of steelhead from hatchery racks, fish traps, weirs, commercial fishing vessels, fish processing plants and angler catches, and after carefully measuring and weighing hundreds of "big" fish, I can state unequivocally that 40 in. does not guarantee 20 lb. To that I will add that I have never weighed a steelhead with a girth of 20 in. or more that did not weigh

The author displaying equipment used to eliminate guesswork about the weight of any steelhead he or his associates caught while angling.

at least 20 lb. Those who contend that their 38 or 39 in. fish weighed 25 lb. or more obviously don't know how disproportionately fat a fish of that length has to be to achieve such a weight. All that aside, if the cost of catch-and-release is never more than a bit of embellishment of steelhead size, I'll live with the heroes of the day who claim they are rewriting the record books.

15

FUTURE CONSIDERATIONS

Over my years of trying to marry steelhead fishing and steelhead management, I've kept watch for pearls of wisdom that show up in the writings and
remarks of others. Whenever I came across something I considered worthy of
reflection I tucked it away in a file I labelled "quotable quotes." As I contemplate what more I might say here, two of those entries stand out. The first was
a comment posted by Dr. Gordon Hartman, a long retired but tireless fisheries
biologist, resource manager and environmental activist, in response to blog
comments made in 2008 by another well-known, strong proponent of responsible resource management, Dr. Robert Lackey:

> It may be the burden of an old person that the farther they can look back
> in life, the farther they may try to look ahead.[27]

The second is part of the signature block of lifelong wild salmon and steelhead advocate Bill Bakke on all his messages sent in support of wild fish and
their habitats:

> One lives with the ghosts of what was and the hunger for what could have
> been.

Those comments would seem to bracket much of what more than half a century of a life with steelhead in British Columbia compels me to say.

It is hard not to be disappointed, frustrated and sometimes even bitter having witnessed the evolution of steelhead fishing in British Columbia over these
past many years. Vancouver Island rivers have fallen victim to the desecration
that inevitably results from the rate and extent of timber harvest visited upon
them, in spite of the best efforts of many good people to avoid such outcomes.
The mainland coast opposite is a steelhead wasteland for all the same reasons.
The early steelhead managers and sport fishers who thought logged watersheds were in recovery mode 60 and 70 years ago couldn't have been more
wrong. The same can be said for their frequently publicized messages that

anglers could do no harm to steelhead populations. The Greater Vancouver area and Lower Fraser Valley have turned to life support from hatcheries that create fisheries distantly removed from those they replaced. Squamish system steelhead hang by threads. The Thompson is a pathetic shadow of its former self. Bella Coola shows no sign of recovery after two decades of steelhead angler exclusion. Who even remembers that the Chilcotin once supported a thriving fishery? And what has become the substitute for lost and forgotten steelhead and cutthroat all over southwestern BC and even beyond? Pink, chum and sockeye salmon, the designated "net species" and the exclusive property of commercial fishermen for the vast majority of the sport fishing history of British Columbia. How many of those who now stand shoulder to shoulder each August in quest of imported pink salmon literally carpeting the Campbell's Sandy Pool know they fish the home water of a legend?

The Dean is still good to those who haven't been there long enough to know otherwise, but the inter-annual variations in run size are troublesome. In northern BC there remains a semblance of what once was, but even there storm clouds loom on the distant horizon. How much more timber harvest, mining development, how many pipelines, IPPs, LNG plants and port developments, how much more commercial fishing? What about the sheer numbers of people moving into small northern communities to partake of all these economic opportunities? How long before all of this conspires to irreversibly alter the remaining rivers and fisheries too many take for granted? Have they already? How many First Nations treaties will preclude access to what remains of our best rivers for those of us who don't qualify or refuse to cough up blood money?

There is no readily reversible factor one can point to that has caused the trends in fish supply clear to those of us who have paid attention. The wave of people occupying and exploiting landscapes and resources is rarely manifested in black and white, but the grey never stops creeping. Someone once said the only problem with the march of progress is it doesn't know when to stop. Does anyone believe that will change? Habitat compromises have been obvious, but fixing broken streams is simply not an option. Less obvious is the fact that, regardless of habitat condition, a given crop of smolts making its way seaward can experience fourfold differences in survival to adult on a regular basis and tenfold differences in extreme cases. The ocean is becoming more inhospitable with each passing decade. Summers are warmer and drier and late summer river flows set new records for low volume and high temperature at a disturbing rate. There is no predictability to most of this, at least none that would be universally accepted. Even if there were, there is little hope for effective or timely management responses to address the circumstances. The supply side of the wild steelhead equation is independent of influence until whatever adults materialize in a given year are in or near their river of origin. Even the patently obvious problem of exploding pinniped populations now conditioned to living in estuaries and rivers and dining on high-value and often impoverished stocks goes unaddressed.

First Nations are the other proverbial wild card with potentially major influence on the supply of steelhead. On one hand they may be the strongest allies in the fight for survival of habitat. On the other, constitutional rights, court decisions and policies place every one of their fishers, commercial netters included, first in line. A growing number of them are prepared to show us just how well they understand that. What avenue is there to deter gillnetting and spearing of winter steelhead in the Cowichan River? Who speaks for what remains of the wild summer steelhead in the Somass system when the gillnets choke the tidal reaches of the river barely upstream from all the gillnets and seines operated by their relatives days before? What about the drift and set nets in the Skeena at Kitwanga, Kitseguecla, Glen Vowell and Kispiox? Who monitors the harvest and where can one find any credible data? Who notices the First Nations anglers who kill as many steelhead as they like wherever and whenever because they can? Short of closing all sport fishing as a demonstrable conservation requirement, governments can't and won't do anything. What will the angling community put on the table in response if it ever gets organized and informed well enough to engage in negotiations with the First Nations over these matters? Why would First Nations even be interested in such an approach? After all, we're just playing with their food.

And what about the demand side? Where are anglers and those authorized to manage them in all of this? Stock assessment they understand. Taking stock of the force that angling has become they don't. Where is the *mea culpa*? What will it take before they understand that there are no more rivers and history is screaming at us that our ability to sustain steelhead populations in these present conditions has no precedent? Who sees the pattern of increasing concentration of effort on the declining number of streams whose fish or fisheries haven't already succumbed? Who is tracking the burgeoning commercial exploitation brought to bear by guiding on every river still alive? Remember, only about 10 per cent of the steelhead rivers in this province are classified and therefore subject to at least some restrictions on guides. Where is the monitoring of angler efficiency and its consequences for fixed or diminishing supplies of fish? How long will we remain anaesthetized to the fact that anglers catching a steadily increasing proportion of a static or declining fish population creates an illusion of abundance? How many boats is too many boats? Commercial fisheries, squarely in anglers' crosshairs for as long as anyone can remember, and often rightfully so, have been dramatically curtailed in terms of the times and places they are permitted to fish and the technology employable. Everyone understands the impact of failure to impose these constraints. Why is sport fishing exempt from parallels?

What countermeasures are offered up by anglers or their regulatory masters? Thankfully we have wild steelhead release regulations throughout the province, in spite of the relentless efforts of a few Neanderthals to reinstate harvest. Winters are milder, later arriving and sooner departing. Steelhead in northern rivers were once protected by ice and snow well in advance of the end of the sport fishing season. Not so today. We can expect more years of more fishing on highly vulnerable overwintering fish populations by both recreational and First Nations fishermen. How much longer will recreational fishery harvest elimination alone be perceived as sufficient to stem the tide? Social media overflows with the mantra of the moment pervading the steelhead sport fishing world. "Keep 'em wet." All good stuff, but does that not disguise the ultimate question? How much catching is too much catching?

Contemporary fly fishers are perhaps the most informed among the greater population of anglers. They love their icons and cite their written legacy regularly. Consider one of the more prominent works of recent times, my friend Trey Combs's *Steelhead Fly Fishing* published in 1991. Flip through that book and compare the rods, reels, lines and flies that were the state of the art at the time with the equipment sported by the average fly fishing enthusiast of today. Even full-time guides can't keep up with the rate of appearance of new and improved rods and lines capable of delivering four-inch, heavily weighted flies distances and depths reachable only by hardware fishermen when Combs's book appeared. When all these anglers add the new era jet boats and inflatables, not to mention helicopters, to their repertoire so they can reach every nook and cranny that competition pushes them to, where does that get us? Given the apparent level of awareness among the fly fishers that now outnumber all other anglers on most of the high-profile wild steelhead rivers of British Columbia, I find their failure to recognize their growing impact disturbing. Patagonia founder Yvon Chouinard had it right:

When all the basketball players are seven feet tall, maybe it's time to raise the hoop.

Those who remain oblivious to outcomes already evident in the southern half of British Columbia need to visit the rivers of the north to learn how quickly our footprint has been imposed and how large it has become. In less than four or five steelhead generations every conceivable opportunity to exploit fish for profit has become the norm. The sensitive and ultra-unproductive tributaries of the Stikine and Taku, the last of the Nass tributaries, the entire mainstem Skeena, the mini-steelhead paradise of the Kluatantan, every lower Skeena tributary large enough to blast a boat through or space for a helicopter to land are all now exploited by commercial interests, naïvely or otherwise pursuing the proverbial grass beyond the mountains. I saw the almost incomprehensible abundance of bull trout in places like the Sustut, the amazing supply of cohabiting resident rainbow and dollies and bull trout in the Nakina, the Nahlin and the Tahltan. I know what happens to those populations following a steady influx of anglers. I fished Chambers Creek, the Kincolith, the Kwinamass, the Ain and Awun, all the Kitlope tributaries and a host of other streams when they were still largely unknown and splendidly isolated from the outwardly altruistic profit seekers of today. I've watched wilderness fishing on the storied "world class" Babine approach the equivalent of a fast food drive-through. Fifteen jet boat passages through the run I chose to begin my day on in 2012 drove the point home. Twenty-five years later I appreciate fully how wise the decision was to exclude guides on gems like the Gitnadoix and Lakelse, and I realize the list of exclusions should have been longer.

Look to the wonderful little Damdochax as a barometer of the evolutionary pathway. I knew it when it was a hand-hewn experience under the careful stewardship of Ken Belford, its original licensee, and Alice, his remarkably capable wife. It was the only classified river in the province regulated for one guide only, in spite of stated policy forbidding such exclusivity. That had everything to do with preservation of an environment and an experience like no other in British Columbia and nothing to do with rewarding someone with a monopoly. In front of me I have copies of Ken's Damdochax/Blackwater angling journals from the 1990s. Their content and context is in stark contrast to the video clips and marketing material extolling the virtues of that petite, gossamer stream and its few hundred fish a mere three steelhead generations later. Here's a quote from Ken following the 1997 season:

I don't know how many other places there are like the Damdochax, but people tell me there are few. In this shrinking world, relationships don't last as long as they used to. But the Damdochax will be there for us in times to come, unroaded and wild and alive.

Maybe it is true nothing is forever but the Damdochax has been there for all time. And it is intact today, just as it always was. Tomorrow is another day and next year will be different too.

Is that not disturbingly similar to the foreword of Fennelly's *Steelhead Paradise* some 35 years earlier?

Consider Mother Skeena herself. Resident opportunity has been supplanted by steelhead purveyors who have become accomplished at manipulating the licensing and administration systems to seize what they treat as their entitlement. No fish is insulated from the one-week thrill seekers who arrive in aircraft and boats at camps and lodges in all those places where steelhead once found refuge from the pressure exerted everywhere else along their journey. Who appreciates that there will be consequences for fishing every steelhead holding run from one end of a season to the other? Who will consider that the more catching that occurs in the Skeena proper, the less available in its tributaries? There are no more fish. There are only more of us with vastly greater knowledge, better equipment and more access than at any time in steelhead history.

Guides and their supporters rise in righteous indignation whenever one hints their activity is not without repercussions. Environmentally friendly economic development is next to godliness and easily sold to those with influence but lacking the knowledge and experience to make informed choices. How many times have I heard the question – well, what do you want, us or industry? When the steelhead guiding industry, in northern BC at least, is increasingly dominated by foreign owners we shouldn't be surprised that it doesn't show up in any meaningful way when fish habitat is on the table. Guides' presence in the forums where commercial fishery impacts on Skeena steelhead or Dean steelhead or Barkley Sound steelhead or Nass steelhead or Taku or Stikine steelhead, the fish so many of them extract a living from or use to subsidize their habit, is equally rare.[28] Contrast that with guides' domination of processes that deal with rod-day allocations or opportunity to discourage or reduce competition by non-resident anglers. Does anyone notice that ever-expanding commercial recreational activity on rivers and ever more constrained commercial fishing activity (none at all in the case of winter steelhead) in the approaches to the same rivers is a tradeoff? The line that once separated the impact of the two fisheries is rapidly becoming blurred. When closure of the last fish cannery in Prince Rupert was announced in late 2015, threatening seasonal employment for dozens if not hundreds of locals, who appreciates their perception that greedy upriver commercial interests are at least partially responsible?

Regardless of the particular river they frequent, do the guides believe their concentration on all the best times and places steelhead can still be found is reasonable? Who understands how minuscule the prerequisites are for the angling guide licence that facilitates this proliferation? Who compares the gap between the plethora of regulations governing hunting guides, their areas of operation and the limits and costs imposed on their comparatively small number of clients with what an angling guide licence facilitates? Limited entry hunting has been a fact of life in British Columbia for decades, for obvious reasons. Are there no similarities between how much hunting is tolerable on a fixed unit of land with finite animal populations versus how much (quality) fishing a given river with a fixed supply of fish can provide? What do angling guides return to the owners for the privilege of disproportionate use of a public resource? How many of them will have to trip over each other on all the waters now exempt from limits on their use before they too cry foul? Where will the rest of us be by then? Does the government agency that now invests responsibility for licensing angling guides in the hands of a centralized digital clearing house, many layers removed from its own staff who are supposed to manage the fisheries, understand the consequences of that step? When will it begin to deliver on its long-standing policy of residents first and its objective of sustaining "quality angling opportunity"?

In *Fisherman's Summer*, first published in 1959, Roderick Haig-Brown spoke to the quality of our sport.

What has to be understood is that the quality of sport is all important. And the quality is not something that can be measured; it is the sum of generations of tradition, ethics and restraint. The quality of the sport is in what anglers themselves have imagined, developed, tested and proved over hundreds of years. It is something that has evolved, not something that has been imposed. It is in what a man dreams of by the fireside at home and goes out next day or next year to try and realize on his favourite lake or stream.

Colleagues (l to r) Dionys DeLeeuw (RIP), Mark Beere and Colin Spence along the Ain River, Haida Gwaii on April 8, 1989.

September 16, 1993. Ken Belford in conversation on the virtues of the Damdochax.

... Let the biologists read and learn their Juliana and Walton and Halford,
their Gordon and Hewitt and LaBranche.... They would know that sport
without limits and restraints, without ethics and tradition, is not sport at
all and can satisfy no thinking person for very long.

Quality angling is no less difficult to define today than it was in 1959. Arriving
at consensus on what it is would itself be a growth industry in the extreme. I
know for certain what it isn't, though. Boats passing through a run 15 times
in a morning, duelling helicopters, a race to be first in at every boat launch,
bigger power plants on bulletproof hulls, a growing parade of entrepreneurs
bent on sucking blood from the steelhead stone, special deals for the favoured
few to buy access denied all others, and on and on. Is there anyone in a position
of power or influence in government office today that begins to understand
these forces and how they conspire to destroy the very fisheries responsible, in
part, for this province's quality of life and its reputation and marketability as a
world class destination? Would they do anything about it if they did? Are there
anglers prepared to demand less in order to sustain a small piece of treasures
once given us?

The voting population of British Columbia is concentrated or, more
properly, sequestered, in the southwestern corner of the province. That population is comprised of a steadily increasing proportion of immigrants and
their descendants from cultures that long ago exhausted the last opportunity
to connect to the natural world. How many of these people do we see on rivers
of the province? I make no judgment on their lifestyles or choice of leisure
pursuits. I merely observe that there is a growing population that doesn't know
and doesn't care about the times, places or issues described in these pages.
How will this play out in years ahead as resource extractors and developers
face off against the environmental protectionists over the places where wild
steelhead linger on? What can we count on from governments dependent on
the electorate of southwestern British Columbia for their hold on power?

Conservation of fishing opportunities obviously begins with protection
of the basic habitat that produces the fish. That in itself has proven to be a
losing proposition for too many watersheds. Nonetheless there are still a few
places where habitat can be allocated for priority use by fish. Surely, in the
vastness of this great province, there is a steelhead stream or two where fish
and fishing are worth prioritizing ahead of timber extraction, mining, water
storage, power production etc. Habitat issues aside, the greatest potential for

doing something other than presiding over the demise of a legacy available nowhere else on the planet resides in how the fisheries of tomorrow are managed. Unlimited access, unlimited commercialism and a technological race to the bottom will inevitably destroy the values still salvageable. If we choose to ignore these realities there will be no purpose in debating what constitutes quality fishing on the once famous rivers of British Columbia.

Some may see my words as hypocritical. After all the times and places I've experienced, many of them courtesy of my former employer, here I am preaching restraint to those who can only dream of such opportunities. I make no apologies for having sampled the bounty of many streams while in-service. Had I not seen and touched those places in honest pursuit of a knowledge base as to what was there and how best to sustain it, I'll dare suggest there wouldn't be classified waters today, nor would there be many of the regulations that have kept wild steelhead hanging on to the extent they have. Critics will be critics, that much I know. And, yes, I too have a jet boat that makes me part of the growing problem on some of the rivers I still choose to visit. But I'll happily leave my boat behind or apply for an equitable opportunity to use it if there could ever be a properly administered program of tailoring the demand for fishing to the supply of something approaching a quality experience.

The author's son. What will his legacy be?

APPENDIX

BRITISH COLUMBIA STEELHEAD SPORT FISHING
REGULATIONS HISTORY

The following is a summary of the more significant regulations brought to bear on steelhead fishing over a period now spanning more than six decades. Not every spatial or temporal detail is included, but enough are here to give a thorough appreciation of the evolutionary pathway and a glimpse into the future. Also included are references to government annual reports that occasionally mentioned angling guides and their predecessor group of licensees once known as small game and angling guides. The time period referenced begins in 1953, the first year when a copy of the regulations was available to the author.

1953
- Roe ban in Vancouver Island District (i.e., Vancouver Island only) and on Capilano, Lynn, Seymour and Cheakamus rivers in Region 2.
- Three-fish daily limit and three days' possession limit on Vancouver Island but only two fish per day in Lower Mainland.
- Forty-fish annual limit on Lower Mainland only. Streams often closed to taking of steelhead March 1.
- Annual fishing licence fees: $2.00 for BC residents, $5.00 for other Canadians, $7.00 for non-Canadians.

1955
- Squamish River added to list of streams where bait was banned.
- Non-BC Canadian annual fishing licence fee reduced to $3.50.

1956
- Prohibition on use of roe on Vancouver Island rescinded.
- Kispiox, Bulkley and Morice rivers in "Northern District" (all territory north of an east/west line across the province at approximately Clinton) added to list of roe ban waters. No explanation given.
- Three fish per day on Lower Mainland district rivers restored but still 40 per year.

1957
- Stawamus River added to list of roe ban waters.

1958
- Notice that a steelhead punch card will be implemented January 1, 1959.

- Season limit of 40 fish in all districts, daily limit of two fish in all districts except the north (still three), possession limit still three days' catch.
- Bulkley River disappears from list of roe-restricted waters.

1965
- Steelhead punch card promised in 1958 regulations booklet finally arrives ($0.25 regardless of angler residence).
- Daily catch limit remains at two, except in northern BC (three per day). Possession limit three days' catch everywhere in the province.

1966
- First detectable reference to "small game and angling guide licence" appears in annual report of the fish and wildlife management agency of the day. The fee was $5.00 and the total revenue generated suggested there would not have been more than 20–25 licences issued province wide. It was unclear how "small game" was defined and how many licensees pursued that versus fishing.
- Steelhead punch card fee for non-Canadians increases to $5.00. Non-Canadian annual fishing licence fee increases to $10.00.

1968
- Province divided into seven regions instead of five districts.
- Province-wide daily catch (harvest) limit became two steelhead and possession limit three daily catch limits.
- Annual fishing licence fee increases to $3.00. Fee discrimination between residents of BC versus other provinces removed.

1970
- Babine River added to roe ban list.

1971
- Steelhead punch card fee rises to $2.00 for any resident of Canada.
- Bulkley River (sections only) added to roe ban list.

1973
- Region 1 (Vancouver Island) reduces summer steelhead daily catch limit to 1 and possession limit to three for the period June 1–October 31. Steelhead in general still two per day and six in possession under all other circumstances throughout province.

1974
- Annual fishing licence fee for any resident of Canada increases to $5.00; non-residents of Canada to $15.00.
- Steelhead punch card fee for residents increases to $3.00, non-Canadian resident fee to $10.00.
- Special Rivers Licence arrives. Non-residents (non-BC and non-Canadian) now require additional $25.00 licence to fish in 13 "exceptional quality" steelhead rivers of the province.
- Steelhead punch card limit becomes 20 fish, but two punch cards per year permissible.
- Region 2 (Lower Mainland) daily steelhead harvest and possession limits fall to one and two fish respectively but only for the smaller, relatively less travelled

streams of the region. All the prominent winter steelhead streams remain at two per day and four in possession.

1975
- Steelhead punch card fee for residents of other provinces of Canada rises to $10.00, the same fee paid by non-Canadians.
- Steelhead punch card limit remains at 20 but only one punch card permissible.
- Summer steelhead limits one per day and three in possession on Vancouver Island June 1–November 30.
- All Lower Mainland steelhead streams 1 per day and 2 in possession.
- Winter angling closures on the Kispiox and Babine rivers. Closure commencement dates vary over time but, generally, take effect December 1.

1976
- Annual fishing licence fee for non-Canadians rises to $15.00.
- Region 1 (Vancouver Island) summer steelhead possession limit reduced to two.
- First ever catch-and-release (C&R) regulations. Five Vancouver Island summer steelhead streams targeted. C&R accompanied by single barbless hook (SBH) but not a bait ban.
- Dean River added to roe ban list.
- Chilcotin/Chilko, Dean possession limits reduced to two and three respectively.
- Skeena Region (Region 6) rivers daily catch and possession limits reduced to two and four respectively.

1977
- Annual steelhead limit remains at 20 but only 10 from any one stream.
- Babine and Kispiox rivers daily catch and possession limits reduced to one and one respectively.

1978
- Region 1 winter steelhead possession limit reduced to four.
- C&R regulations imposed on most Region 1 summer steelhead streams.
- Several new Skeena tributaries added to one daily and one or two possession limit list.
- Zymoetz and Suskwa rivers added to roe ban list.
- Chilcotin/Chilko rivers daily limit reduced to one (still two possession but also two annual).
- Dean River becomes first in the province where use of any natural bait prohibited.

1979
- Thompson daily limit becomes one, possession limit one or two depending on time of season.
- In-season Emergency Order mandating C&R of all wild steelhead catch on Vancouver Island, winter 1979–80.

1980
- "Wild Steelhead Release" and "Steelhead Release" regulations appeared in the regulations booklet for the first time. Both carry SBH and bait ban requirements.
- Region 1 goes to wild steelhead daily limit of one, possession two but aggregate steelhead (i.e., wild plus hatchery) limit of two daily and four possession.

Additional limit of five wild steelhead annually (still 10 aggregate) but no more than two wild in any calendar month.
- Region 2 adopts daily limit of one steelhead (possession two), five wild steelhead per year but no more than five from one stream and two wild per month.
- Regions 3 (Thompson), 5 (Chilcotin) and 6 (Skeena) still at 20 wild steelhead per year but no more than 10 from any one stream, and some more restrictive stream specific quotas prevail.
- Thompson closure extended from March 31–May 31 to January 1–May 31.
- Bulkley roe ban extended.

1981
- Province-wide annual quota of 10 steelhead.
- Queen Charlottes added to area where only five steelhead could be taken annually from any one stream.
- Annual limit of two for Kispiox and Babine rivers.
- Steelhead release on Kispiox River August 15–September 30.
- The Regulations Synopsis noted the province was considering a province-wide roe ban in all non-tidal waters and a single-hook restriction in all non-tidal streams.

1982
- Annual fishing licence fee increases for all licence classes. Residents fee $13.00, non-resident Canadian fee $15.00 and non-Canadian fee $23.00. All basic licence fees include a $3.00 surcharge for newly implemented Habitat Conservation Fund.
- Steelhead licence fee increases for residents ($6.00). Fees for non-BC Canadians and non-Canadians both $15.00.
- Regulations booklet noted that over 200 written and oral responses had been received regarding the roe ban, single-hook proposal. Decision to maintain site-specific regulations only.
- Mandatory release of summer steelhead, SBH and bait ban ubiquitous on Vancouver Island. Wild winter steelhead mandatory release December 1–March 1.
- Some Region 2 streams appear on December 1–March 31 C&R list (Squamish, Cheakamus, Capilano and Seymour but not Lynn).

1983
- Most Lower Mainland (Region 2) streams added to wild steelhead release list December 1–April 30. (Squamish streams to May 31.)
- SBH ubiquitous in Region 2 except Chilliwack.
- Government annual reports indicate the number of small game and angling guide licences issued that year totalled 161. Licence fee $15.00.

1984
- "Steelhead Designated Water" definition appears in regulations synopsis. Anglers fishing waters so designated must cease fishing for any species of fish once a daily limit of steelhead has been caught and retained. Notice given that this rule will apply to all waters, not just those designated, as of 1985.
- Thompson River steelhead harvest limits reduced to one per day, two per month (five per year from the Fraser River at Lytton).
- Winter closures (December 1–June 30) standardized for most (not all) prominent Skeena tributaries.

1985

- Steelhead Designated Water regulation applicable to specifically named rivers replaced by general restriction requiring anglers to cease fishing for any species in any water if they have caught and retained a daily limit of steelhead that day.
- Region 1 imposes wild steelhead release all year, SBH on all streams all year, bait ban on all streams May 1–November 30.
- Region 2 imposes wild steelhead release, SBH on all streams all year.
- Separate regulations governing the use of "bait" versus "roe" combined into a single rule banning all bait. Much controversy over the legal definition of bait.
- Special Rivers Licence becomes Special Waters Licence. Fee unchanged.

1987

- Special Rivers Licence discontinued in favour of "Special Fisheries Permit." Applicable fees are $10.00 for residents of BC, $25.00 for residents of other provinces and $150.00 for non-Canadians, but latter only applicable to Dean River.
- Steelhead licence fee increases for non-BC Canadians ($15.00) and non-Canadians ($40.00).
- Permit system for non-resident anglers commenced on the Dean River.
- Annual government reports make specific mention of angling guide and assistant angling guide licences for the first time but no breakdown provided. Report notes "guiding was becoming a significant economic activity, particularly in northwestern British Columbia."

1989

- Annual angling licence fee increases. BC residents to $15.00, other Canadians to $17.00, non-Canadians to $25.00.
- Steelhead licence fee increases also. Residents to $7.00, other Canadians to $17.00, non-Canadians to $42.00.
- Thompson River added to the growing list of wild steelhead C&R waters.
- In-season emergency order requiring release of all wild steelhead in the Skeena system, which had otherwise already been reduced to one per year.
- All summer steelhead tributaries of Skeena and Nass closed December 1–June 30.

1990

- Annual angling licence fee increases. BC residents to $17.00, other Canadians to $19.00, non-Canadians to $27.00.
- Classified waters regulations take effect April 1. Non-resident Canadians and non-resident aliens required to pay daily fees of $10.00 and $20.00 to fish Class 2 and Class 1 waters respectively. Residents pay $1.00 per day on Class 1 waters only. Classified waters list includes five class 1 and 40 class 2.
- Number of angling guide licences and guided rod-day quotas on classified rivers, most of which are in the Skeena watershed, frozen at pre-existing levels.
- Angling guides pay $100.00 licence fee plus $1.00 for each rod day included in their quota. Assistant angling guide licence fee $25.00.
- Province-wide numbers of angling guides and assistant angling guides licensed were 318 and 190 respectively.
- Both the Skeena and Nass systems C&R only for wild summer steelhead November 1–December 31 and closed thereafter. (The 1990 fishing season on the Skeena marked the first year in which the annual quota for steelhead was one. Severe conservation concerns that persisted from 1991 through 1996 inclusive

forced in-season measures reducing the quota to zero in each of those years. In 1997 the in-season approach was abandoned in favour of blanket C&R regulations for the entire Skeena and Nass watersheds right from the start of the season.)

1992
- Wild steelhead catch and possession limits on the Dean and Chilko/Chilcotin rivers reduced to one per year.
- All Skeena and Nass tributaries included in the growing list of waters prohibiting the use of bait.

1994
- Annual angling licence fee increases. BC residents to $23.00, other Canadians to $28.00, non-Canadians to $40.00.
- Annual steelhead licence fees adjusted. BC residents fee rises to $10.00, other Canadians' fee rises to $20.00, but non-Canadians' fee falls to $20.00.
- Estimated number of angling guide and assistant angling guide licences listed as 330 each.
- Angling guide and assistant angling guide licence fees rise to $200.00 and $45.00 respectively.

1996
- Annual steelhead licence fee increases for both non-BC Canadians and non-Canadians. Both now pay $30.00.
- Dean, Chilko/Chilcotin and Bella Coola rivers added to the list of wild steelhead C&R waters.

1997
- Annual angling licence fee increases. BC resident fee rises to $30.00, other Canadians to $40.00 and non-Canadians to $55.00
- Annual steelhead angling licence fee increases. BC resident fee rises to $15.00, other Canadians and non-Canadians to $40.00.
- Annual Classified Waters Licence for residents implemented ($10.00).
- (As above) Skeena and Nass watersheds identified before the fishing season begins as wild steelhead C&R-only waters. Up front announcement precludes the necessity for in-season emergency orders imposed annually since 1988.

1998
- Region 5 adopts wide wild steelhead C&R for all streams.
- Bella Coola River closed to steelhead angling in attempt to persuade First Nation to stop gillnetting steelhead if sport fishery remains as wild steelhead C&R.
- Angling guide and assistant angling guide licence fees rise to $240.00 and $100.00 respectively.

1999
- Emergency in-season angling closures imposed on numerous Vancouver Island steelhead rivers.
- SBH mandatory on all streams of Regions 1, 2, 3, 5 and all Region 6 streams draining into the Pacific Ocean.

2001
- Use of bait prohibited on all Vancouver Island streams all year.

2003

- Annual angling licence fee increases. BC resident fee rises to $36.00, other Canadians to $55.00 and non-Canadians to $80.00.
- Annual steelhead angling stamp fee increases. BC resident fee rises to $25.00, other Canadians and non-Canadians to $60.00.
- Daily licence fees for non-residents and non-resident aliens to fish classified waters rise to $20.00 and $40.00 for Class 2 and Class 1 waters respectively.
- Annual licence fee for BC residents to fish classified waters rises to $15.00.

2004

- Angling guide and assistant angling guide licence fees rise to $450.00 and $150.00 respectively.

2006

- Guided rod day fees on classified waters increase from $10.00 per day to $21.00 per day and $26.00 per day for Class 2 and Class 1 waters respectively.

2007

- Province-wide wild steelhead release regulation in effect April 1.
- SBH for all river fisheries mandatory.
- Some angling closures imposed on Vancouver Island in 1999 partially rescinded. Lowermost reaches of several east coast rivers (Chemainus, Cluxewe, Englishman, Little Qualicum, Nanaimo, Oyster, Puntledge, Quatse and Quinsam) reopened to C&R-only, SBH and artificial lure only.
- Guided rod day fees on classified waters increase to $26.00 per day and $31.00 per day for Class 2 and Class 1 waters respectively.

2011

- Regulations governing times when non-residents of Canada can fish on some classified steelhead tributaries of the Skeena system and sections of the Skeena River itself adjusted to give resident anglers exclusive access on weekends.
- Thompson River closed to fishing September 30. Shotgun opening possibility if in-season escapement estimates warrant.

2014

- Thompson River regulations amended from closed to angling September 30 to open unless in-season escapement estimates warrant closure after that date.
- Long-standing debate around terminal gear reaches decision point with imposition of bait ban.

2015

- Process for issuing angling guide and assistant angling guide licences distantly removed from regional offices and regions where guiding occurs.
- Provincial fisheries staff no longer able to monitor guided angler activity to the extent necessary to respond to basic questions such as how many guides and assistants operate and how much activity they bring to bear on specific (unclassified) waters.

NOTES

1   Kennedy Warne, "Blue Haven: New Zealand Marine Reserves Are a Model for the World," *National Geographic*, April 2007.

2   Francis C. Whitehorse, *Sport Fishes of Western Canada and Some Others* (Toronto: McClelland and Stewart, 1946).

3   Van Gorman Egan, *Rivers on My Mind* (Campbell River, BC: Riverside Publications, 1998).

4   I note that Lee Straight published an article in the *Vancouver Sun* on April 27, 1968 describing an inaugural jet boat trip on the Squamish one week earlier. Fair enough, but I never encountered another power boat of any description on the river on days I spent there.

5   December returning steelhead were once common in many Vancouver Island streams. They are rare to non-existent today.

6   The award conferred by the 92-year-old, world famous Tyee Club of Campbell River to any member who catches a chinook salmon weighing 60 lb. or more. There has been one such fish in the past 35 years.

7   Champoeg Press, Forest Grove, Ore.

8   Ibid., 73.

9   Edward Weeks, *Fresh Waters* (Toronto: Little, Brown, 1968).

10 BC fly fishing history buff Art Lingren has noted that Egan's Run is actually named after Maxine rather than Van.

11 Don McCulloch, a former professional colleague (DFO Fisheries Officer in Masset), caught what may be the largest ever winter steelhead recorded in British Columbia. Don's 33 lb. fish, taken in the Yakoun on November 16, 1972, measured 42 in. long by 25.5 in. girth.

12 Richardson, *Lee Richardson's B.C.*, 43.

13 A stunning development occurred after this chapter was written. On October 25, 2016, after nine years of interaction with multiple interest groups dedicated to rebuilding and/or redesigning the dam and diversion works to better accommodate fish passage, BC Hydro announced it intends to decommission the entire operation, hopefully in 2017.

14 An unpublished document by Department of Fisheries and Oceans staff (Hyatt and Steer), dated 1987, stated that prior to the 1954 upgrade to the original vertical-slot fishway constructed in 1927, it was believed salmon, with the exception of summer steelhead, were restricted largely to habitats below Stamp Falls.

15 The figure of 20 was quoted by Judge Joe according to sworn testimony of the accused.

16 The best available records indicate the guides accounted for half the total catch of steelhead on the SSS system by 2000 and likely an even higher proportion in the years that followed.

17 I'm told that non-guides now run jet boats on the system to challenge the preferential treatment of guides. Enforcement authorities are turning a blind eye.

18   The first ever trip to the Nilkitkwa confluence was described in detail in one of a series of articles about the Babine written by Ed Neal, the outdoors columnist of the day for the *San Francisco News*. Ed Neal was kind enough to provide photocopies of all his articles and photographs of his fishing and hunting adventures to the Babine, Sustut and Bear Lake areas in 1955 and 1956.

19   There is only one small spot on the 13 miles of water accessible by jet boat from the fisheries weir where there is any view of a distant mountain peak, and then only if no fog or cloud is hanging over the river. It is a far more impressive scenic environment along the section of river where the lower lodges are located.

20   Today I can pull up a website and look at the historic stream-flow records for the Babine. Those data cover 43 years and confirm that the river discharge in September and most of October in 1976 was the highest ever measured at that time of year.

21   For records' sake, the dimensions of that fish were 41.5 in. length and 23.25 in. girth. I'll have more to say about length, girth and corresponding weights of steelhead in a later chapter.

22   I have requested the report on what transpired at Moricetown in 2013 on four different occasions, the last being in late 2014. Nothing has been forthcoming.

23   Sockeye were not yet legal quarry, so gear choice never focused on them and none were ever caught on our standard bar fishing rigs.

24   "The Tragedy of the Commons" is a milestone article written by Garrett Hardin and published in *Science* in December 1968. Between Hardin and Haig-Brown, wiser words around resource-management issues have yet to be spoken.

25   Three thousand spawning steelhead were counted in the Bear River in the spring of 1971. Bear River once supported the single largest spawning population of chinook salmon in the Skeena system. That number is dramatically lower today.

26   I described the Tyee test fishery in my previous book, along with other important details around the August 1, 1998, occurrence and the eventual fate of those fish.

27   I've since seen a similar comment attributed to Winston Churchill although no details were recorded.

28   To be fair, one guide (Keith Douglas) has done more than his share to keep the commercial fishery steelhead interception issue in the forefront.

SELECTED REFERENCES

Alevras, John. *Leaves from a Steelheader's Diary.* Portland, Ore.: Frank Amato Publications, 2010.

Combs, Trey. *Steelhead Fly Fishing.* New York: Lyons and Buford, 1991.

Egan, Van Gorman. *Rivers on My Mind.* Campbell River, BC: Riverside Publications, 1998.

Fennelly, John F. *Steelhead Paradise.* Vancouver: Mitchell Press, 1963.

Haig-Brown, Roderick L. *Fisherman's Summer.* New York: William Morrow and Co., 1959.

Hardin, Garrett. "The Tragedy of the Commons." *Science* 162, no. 3859 (1968): 1243–48.

Hooton, R.S. *Skeena Steelhead – Unknown Past, Uncertain Future.* Portland, Ore.: Frank Amato Publications, 2012.

Richardson, Lee. *Lee Richardson's B.C. – Tales of Fishing in British Columbia.* Forest Grove, Ore: Champoeg Press, 1978.

Taylor, Ben. *I Know Bill Schaadt – Portrait of a Fishing Legend.* Kenwood, Calif.: Benjamin R. Taylor, 2013.

Warne, Kennedy. "Blue Haven: New Zealand Marine Reserves Are a Model for the World." *National Geographic,* April 2007.

Weeks, Edward. *Fresh Waters.* Toronto: Little, Brown, 1968.

Whitehorse, Francis C. *Sport Fishes of Western Canada and Some Others.* Toronto: McClelland and Stewart, 1946.

Wulff, Lee. "Backcasts." *Trout Magazine,* Spring 1987. Arlington, Va.: Trout Unlimited.

Wahl, Ralph. *One Man's Steelhead Shangri-La.* Portland, Ore.: Frank Amato Publications, 1989.

EXTENDED PHOTO CAPTIONS

Page 28. Tools of the trade circa 1970s. A rare 4 in. solid face Dural Major Hardy reel mounted on a model 1263 Fenwick 10½ ft. fibreglass rod. The reel was purchased from Cecil Brown, one of the original owners of Norlakes Lodge on Babine Lake, the forerunner of Babine Norlakes'Steelhead Lodge on the upper Babine River.

Page 32. The 19.5 lb. fish from Wilson's Riffle, April 3, 1972. The rod, an 11 ft. Richmake from Harkley and Haywood, the reel a John Milner customized Silex Superba.

Page 33 (top). May 2, 1975. Pulling on one on the Squamish one run downstream and opposite from where the Cheakamus entered that year. My Zodiac Cadet and 10 HP Mercury provided the access.

Page 33 (bottom). The end result, 22 lb., my largest ever Squamish fish and the only coloured fish I ever caught in that river. Photo by Jim Penner.

Page 37. Prime water conditions in the gravel pit stretch, mid-winter, 1972. Photo by Ted Harding Jr.

Page 40-41. The April 2014 fish that inspired both memories and hope. (The fish is completely immersed in water, as evidenced by the bubbles appearing above and to the left of its eye.)

Page 47. Friend and university classmate Ric Olmsted (RIP) about to beach a large male at the "Hole in the Bend" run, Englishman River, as the deep freeze set in, December 28, 1978. The river's edge below him is already frozen and slush ice was thickening midstream.

Page 49 (top). Grassy Banks run at a less than ideal fishing flow in late January 1979. The worst of the cold weather that had beset the area between Christmas and New Year's was past.

Page 49 (middle). Grassy Banks in January 1981. The February 25, 1979, flood effects were becoming increasingly obvious.

Page 49 (bottom). Grassy Banks in March 1983. The changes continued to unfold. Comparative pictures became impossible after the mid-1980s as the naturally regenerating alder, maple and fir trees in the foreground grew to the point of obscuring the view.

Page 50 (top). The site known as "Top Bridge." Grassy Banks is just out of sight upstream. The old concrete bridge abutments provide an excellent reference point to appreciate the severity of the floods that befell the Englishman between 1979 and 1983.

Page 50 (middle). The old bridge abutment on river left near the peak of the flow event that occurred on February 24-25, 1979. There are no precise flow records for the date but a reasonable estimate of the flow at the time of the photo was 450 cubic metres per second (roughly 16,000 cubic feet per second).

Page 50 (bottom). The same view as the top image a few hours after the peak of the February 24-25, 1979 flood.

Page 51 (top). The upstream view from the twin cedars at river's edge in front of our home during the cold snap that descended between Christmas 1978 and New Year's. These trees had clearly been firmly anchored to the bank for many decades.

Page 51 (bottom). The downstream view from the same twin cedars showing the platform that I had first encountered near the end of a day's fishing with Ted Harding in 1971. The outboard half of the twin spire was torn away and washed seaward by the major flood of February 24-25, 1979. The remaining half is now gone as well, although its date of departure is unknown. The adjacent stream bank that I was convinced was impenetrable when we lived there now appears vulnerable.

Page 52 (top). The April 1979 beach seining operation that captured 126 steelhead in one set at the Claybanks. The major changes in this stretch of the river following the February 24–25 flood that year had only just begun.

Page 52 (bottom). Some of the 126 resting quietly in the seine net before being sorted and tagged.

Page 53. The Claybanks Run, or what remains of it, February 5, 2012. The picture was taken from precisely the same spot as Fig. 18.

Page 67. Day one at "The Circus," January 5, 1983, and co-worker Maurice Lirette playing a fish. The jet boat is parked at river left just inside the turbulent chute leaving the run.

Page 70. The author in his 10 HP prop-driven Zodiac running the tail out of a Gold River canyon pool slightly over two miles upstream from tidewater on a perfect fishing flow in June 1974. Note the size and porosity of the substrate. Dogwoods bloom in the background. They too seem to have disappeared. Photo by Ted Harding Jr.

Page 74 (first). May 18, 1974. Ted Harding fishing a magnificent piece of water near the upstream end of tidal influence. We referred to this run as Tidewater Flats. The river's substrate at the time was uniformly clean, porous cobble from water's edge behind Ted all the way across the run to the rock bluff from which I took the picture.

Page 74 (second). April 8, 1983. The same piece of water eight years later. The infilling of the substrate with finer material was well underway, as was the deposition of sand and gravel at river's edge behind the angler.

Page 74 (third). November 2, 2003 (high tide, strong wind). Twenty years later the same piece of water is dramatically different. The wind-rippled surface hides the extent of deposition of fine material all down the centre of the run.

Page 74 (fourth). June 4, 2013. Another decade and the configuration of the run has changed dramatically again.

Page 75 (top). Late June 1973. A 16+ lb. summer steelhead taken at the Sewage outfall run and pictured at the pullout on the pulp mill road immediately downstream from the Big Bend pool. The pullout was all but washed away on November 13, 1975 and had to be rebuilt. Note the alder at water's edge on river left, the three tallest trees in the background and the average size of the river substrate. Photographer unknown.

Page 75 (bottom). March 26, 2012. The same view 39 years later. The three trees are gone and the substrate is well subscribed with finer material.

Page 76 (top). May 23, 1976. The perch from which my 22+ lb. summer run steelhead was taken two days after this photo. Again, note the dogwood trees on the far bank. They are absent today. Photo by Doug Morrison.

Page 76 (bottom). March 26, 2012. Thirty-six years later the same spot appears sterile. The three large boulders are the same ones behind me, just upstream from the rock I am perched on in Fig. 32, although they have been slightly rearranged. The river discharge in the May 23, 1976 photo is roughly double that on March 26, 2012, but that does not begin to account for the difference over time.

Page 77 (top). May 18, 1974. The downstream view from the rock bluff on river left at the run I referred to as Tidewater Flats.

Page 77 (middle). June 4, 2003. The same view from the same location 29 years later.

Page 77 (bottom). June 4, 2013. Same place, same view after another decade.

Page 78 (top). December 23, 1974. Looking downstream from the tail out of the Lake Pool which we had accessed with the Zodiac and 10 HP prop motor that day. The cascading water that was diverted to address icing problems on the pulp mill road but destroyed the trail to Timmons' Run can be seen in the background at top left.

Page 78-79 (bottom). March 26, 2012. The same view as Fig. 37. The holding water that once produced steelhead consistently was completely rearranged, vastly reduced in area and gave that same impression of sterility. Wading these areas is much easier today because the substrate is flood scoured often enough to prevent the growth of grease like diatoms that once coated every cobble.

Page 79 (right). The author playing a fish from a casting perch on river left opposite the pulp mill water intake works, March 29, 1974 (photo by Gerry Taylor). The pulp mill road is clearly visible in the background. There were four or five large old growth cedar stumps right at water's edge along that side of the river. The one pictured served as a perfect position to drift a cast through the best of the holding water along that reach, not more than a rod's length from the base of the stumps. They all bore the spring board notches used by fallers in earlier days. The trees had obviously been felled directly into the river and floated out. Following the November 13, 1975 flood, that age-old stump and all the others were gone and the formerly prime holding and fishing water on that side of the river replaced by a beach comprised of silt, sand and fine gravel. Two of the larger fish of the many hooked from atop that stump that day appear in the photos below. Friend Ian Carruthers (RIP) hoists a 20 lb. male (left), myself a 16 lb. female (middle). Latter photo by Gerry Taylor.

Page 80 (top). November 13, 1975. This picture was taken from the bridge crossing on the Heber River. The flooded baseball park and track can be seen in the left centre background. Gold River is raging from right to left in the distant background. Near the centre of the photo the Heber is lapping at the underside of the suspension bridge between the local school and the cinder track and playing fields.

Page 80 (bottom). The same view as above image on March 21, 2014. The trees along river left have obviously grown to the point where the track and playing fields are almost completely obscured. The vertical distance between the bridge and the river surface below is at least 15 feet at this average Heber River flow.

Page 81 (top). The Big Bend on the Gold River pulp mill road near the peak of the November 13, 1975 flood. The water was surging over the road and hitting the rock face on the right hand side with such force that there were large back eddies forming on both the upstream and downstream sides of the rock face. The logs in the foreground were circling around the upstream eddy. Note the square ends on the larger logs. How far downriver had they travelled before accumulating at that point?

Page 81 (middle). Another view of the pulp mill road flooding at the Big Bend on November 13, 1975. This photo was taken approximately one hour after the photo above. The peak of the flood was over and the river was beginning to drop.

Page 81 (bottom). The same view as the middle image, several months after the flood event and road repairs. The bar in the centre of the river was never a feature of the river before the flood.

Page 82 (top). The Number 2 Bridge on the mainline logging road immediately upstream of the town of Gold River shortly after the peak flow on November 13, 1975.

Page 82 (bottom). The Number 2 bridge at an average late winter flow a few months later.

Page 87 (top). Typical stream bank condition immediately adjacent to the Salmon River–Memekay River confluence as experienced in March 1980.

Page 87 (bottom). River right at the Salmon/Memekay confluence area on January 13, 1981. The tail of the confluence pool appears at approximately 0900.

Page 88 (top). Aerial view looking downstream at the Salmon–Memekay confluence on March 27, 1986. The confluence pool is in the centre of the photo with the Salmon flowing from bottom right to top left. The Memekay flows from left to right from about 0800. The two preceding ground level photos were taken from the right bank of the Salmon at the confluence pool. The entire area was utterly unrecognizable relative to the preceding two photos from five and six years earlier.

Page 88 (middle). Aerial view looking downstream at the Salmon–Memekay confluence on October 24, 2002. The Salmon is flowing from approximately 0400, the Memekay from bottom centre with the confluence mid frame, slightly left of centre. The tremendous accumulations of logging-related debris evident from the 1980 and 1981 photos had long since been flushed from the system, leaving behind a biological wasteland of fine material continually rearranged by freshets.

Page 88 (bottom). More biological wasteland. An aerial view of the Salmon looking downstream near the Big Tree Creek confluence on October 24, 2002. Big Tree Creek is the thin line from 0500 to join the Salmon mid-frame slightly right of centre.

Page 90. Logging our way along the road to the launch site at the Pallan's Bridge washout, January 22, 1981. Logs had been strewn across the road during a recent freshet and the road had become a secondary stream channel that ran for several hundred metres across the floodplain.

Page 91. Clearing a navigation channel on the Salmon between the confluences of Big Tree Creek and Memekay River, January 23, 1981.

Page 92. Typical launch and retrieval circumstances. Pulling the latest edition of the brood stock capture machine out of the Salmon at the Pallan's Bridge washout. We didn't even have the benefit of a four-wheel drive for any of the years we operated there.

Page 94. Serious business. Salvaging gear from the 16 ft. Smokercraft plastered against the logs after failing to negotiate a hairpin turn halfway between the Memekay confluence and the Big Tree confluence on February 22, 1984.

Page 99. The 29.3 lb. fish caught by Cyril Webster (top), local fish and game club brood stock collection volunteer, on March 18, 1979, along with other examples of large steelhead.

Page 104 (top). The Catherine Creek confluence pool as it appeared in April 1975. This piece of water remained essentially unaltered from my first visit that year until I last saw it a decade later.

Page 104 (bottom). The obstruction that was considered to be a barrier to winter steelhead passage to the upper Tsitika River as of April 1975. It was located about one mile downstream from the Catherine Creek confluence. The worst of the cascade was actually just out of the picture frame on the downstream side. Rick Axford is exploring every possible nook and cranny around the obstruction in search of winter steelhead.

Page 106. A bag of fish coming over the stern roller of Miss Cyanea, the smallest of the seine vessels present at the time, off the tip of Growler Cove on July 11, 1978. Ian Carruthers, an ardent steelheader, salmon fishing guide and friend of the author, was a crew member on this vessel

Page 109 (top). The adult brood stock transport vessel specifically designed to ferry the Tsitika adults from capture sites on the river to the waiting tank truck at the nearest road access.

Page 109 (bottom). On their way, November 5, 1982.

Page 111. August 16, 1977. The day to remember. There were ten fish present, although only nine can be seen in this frame. There were at least a dozen juvenile steelhead of three different age classes hovering about their adult relatives. The quality of the substrate in this run and elsewhere along the river stood testament to what prime steelhead habitat looked like in an undisturbed watershed.

Page 124. Returning downriver over Papermill Dam, Somass River, in my recently purchased Valco jet boat, November 27, 1983. This was the only non-government jet boat on the river at the time. Rob Hobby photo.

Page 125 (top). Long-time friend Rob Hobby with the first hatchery steelhead caught on our inaugural jet boat excursion on the Somass River, November 27, 1983.

Page 125 (bottom left). Senior Conservation Officer John Merriman with a brace of hatchery clones, Stamp River, January 22, 1984.

Page 125 (bottom right). My father with our limit of hatchery fish, Somass River, December 17, 1985.

Page 126 (top). The First Nations fishery in full bloom on the Somass River, June 7, 2016.

Page 126 (bottom). Bank to bank gill net being drifted down the lower Somass on the incoming tide, June 7, 2016.

Page 128-129. Proceeds of a carefully hidden sunken gill net pulled from the Somass River by the author on November 27, 1983. The five prime winter steelhead tangled in the net were turned over to the Port Alberni Conservation Officers for disposal.

Page 135. The only run ("log jam") anywhere along the upper 15 miles of the Babine River that affords a view of distant mountains.

Page 137. An aerial view of Triple Header Lodge on the lower Babine River in September 2007. The views from river's edge are far more attractive than seen on the upper river. Photo courtesy of Rex Lester.

Page 139. The only other non-guide operated jet boat I ever encountered on the Babine River in at least the first ten years I visited it. Three heavyweights in a tinfoil john boat with a 50 HP motor was a death wish in the making. They had serious rock encounters more than once.

Page 141. My son Brock, age 11 at the time, with the 38 in. fish that attacked the #10 deer hair pattern.

Page 143. The buck bug gets introduced to Skeena steelhead. Bulkley River, September 26, 1990. Oregon Bob Hooton photo.

Page 144. That first ever Babine steelhead taken on a buck bug, September 29, 1990, and Oregon Bob's photo that the BC Federation of Fly Fishers used on its membership solicitation brochures.

Page 150 (top). The operculum hooked fish that necessitated the boat chase, October 2, 2001. Photo by Steve Pettit.

Page 150 (bottom). Many of the 1998 fish, especially the larger males, were uncharacteristically fat, although not many displayed quite the condition factor evident here. This one too was victimized by the bug. Photo by Steve Pettit, September 29, 1998.

Page 153. My career best, a legitimate 30 lb. fish, September 6, 1987. Photo by Lori Hooton.

Page 154-155. Five minutes from our home to the river, a quick boat launch and five more minutes down the river to this.

Page 159. The Driftwood Canyon rapid that demanded the portage on November 29, 1987. The worst of the water is just beyond the left margin of the photo. Photo by Steve Pettit.

Page 160 (top). The first of many. John Taylor netting Pete Broomhall's fish that couldn't resist Art Lingren's Black Practitioner in the newly discovered bucket, September 23, 1993.

Page 160 (bottom). Brock at age 15 with one of the better fish to come from the magic little bedrock bathtub that served as a remarkable barometer of what was in and moving through the river. This one was an honest 15 lb. and led us on a long chase down the river on September 6, 1998.

Page 168-169. The dip netting site at the upstream end of Moricetown Canyon during the height of the annual steelhead population estimation program, September 10, 2010. Follow the sequence on page 169 for the evidence of science in action. More than 6,300 steelhead were handled this way that year.

Page 170 (top). All too typical net scars, abrasions and split fins displayed by a fish that had been tagged by dip netters at Moricetown and recaptured 11 miles upstream several weeks later – with all the energy of a wet rag. Photo by Steve Pettit.

Page 170 (bottom). A Moricetown tagged steelhead found and photographed by Keith Douglas on or about September 10, 2013 approximately 21 miles upstream from Moricetown.

Page 171. A near-dead Moricetown tagged steelhead photographed by the author approximately halfway between Moricetown and Smithers in September 2013.

Page 175. The author's children and father on a lower Skeena bar, early August 1988.

Page 176. The decaying proceeds of a First Nations set gill net left unattended in the Skeena River near Glen Vowell. The net was pulled by DFO fisheries officers on September 24, 2012 after complaints from local anglers. It contained 41 steelhead, 1 coho and 2 chum salmon. There were no consequences for this colossal waste. Photo by Keith Beverly.

Page 178. Brock, age 5, weighing in his first steelhead. Skeena River, September 18, 1988.

Page 181 (top). Brock, age 7, with a magnificent 24 lb. steelhead he hooked while fishing from the boat on October 8, 1990 in the run where Bob York and I had our streamside chat days before.

Page 181 (bottom). My father with another beautiful 20+ lb. plus fish from the big water near the mouth of the Kispiox, September17, 1989.

Page 184. "Where's the rest of it?" Reaching for the tape to measure the hour and 40 minute fish, September 14, 2010. Photo by Dana Atagi.

Page 186. A buck bug and dry line fish caught by the author at "Obvious," October 4, 1990 following the two Bobs drift trip on the Babine.

Page 190. The moose encounter on the Kitimat River in late March 1987. The photo was taken just as the moose decided it was time to give up on entertaining us and return to the bush from whence it came. Photo by George Schultze.

Page 192. Friend and colleague Colin Spence with one of 48, Ishkheenickh River, May 12, 1989.

Page 194 (top). First Nations gill net strung across one of the two main intertidal channels of the lower Ishkheenickh at low tide on April 12, 1991. There was another net similarly deployed in the other channel. The net pictured contained three prime female steelhead, the other two males

Page 194 (bottom). The Nass–Ishkheenickh confluence area as it appeared at roughly the same time as our late April adventure in 1991. The side channel we had to use to gain access to the river upstream from the logs is hidden in the trees beyond the left margin of the photo. Its entry point to the Nass was just beyond the top left corner of the photo. The nets observed above were set in each of the two channels above the logs.

Page 196 (top). Overcoming the first obstacle on the journey through the side channel and maze of logs on the lower Ishkheenickh on April 27, 1991

Page 196 (bottom). The next obstacle. This was the point where we would be required to make the right angle turn to enter the boat-width channel coming downstream on the return trip.

Page 198. Cress Farrow (right) and Gerry Taylor engaged in conversation at the uppermost run Gerry and I had hiked to on April 29, 1991.

Page 199 (top). Gerry with one of many landed from the run we'd all but finished with when Cress Farrow arrived in his hovercraft on April 29, 1991. This was arguably the most attractive and productive run on the entire Ishkheenickh River at the time.

Page 199 (bottom). Looking back upstream at Gerry Taylor standing atop the spruce tree that Cress Farrow and friends had cut the picket fence of branches off and another one that they took a section out of to facilitate passage down into the Nass River on April 30.

Page 201 (left). The wall of ice and snow deposited on the lower Ishkheenickh sometime prior to our visit in late April 1991. The ice can be seen on the right bank well above the base of the streamside alders, some of which had been broken and/or partially de-barked by the force of the event.

Page 201 (right). Gerry holding a cobble pried from the wall of snow and ice behind him. His 10½ ft. rod is propped against the wall.

Page 203 (top). Brock with a typical lower Ishkheenickh fish he caught on April 25, 1992.

Page 203 (bottom). Brock, bike and inflatable shortly after arrival at the campsite run on the Ishkheenickh, April 26, 1992.

Page 208. Steelhead lined up in regimental fashion at the outlet of Johanson Lake on September 30, 1986.

Page 210 (top). Fisheries technicians Ron Tetreau (left) and Sig Hatlevik (right) and their helicopter pilot with a triple header while tagging steelhead in the upper Sustut River a few hundred yards downstream from the Johanson River confluence on September 30, 1986. The river was at a typical flow stage for that time of year and obviously small.

Page 210 (bottom). Fisheries technician Ron Tetreau partaking of steelhead tagging on the Sustut River about half a mile downstream from the Bear River confluence on September 28, 1986. The sign on the rock requests anglers to report tagged steelhead. In three days in that immediate vicinity, by far the most productive on the lower Sustut River, Ron and the author did not see another human being.

Page 211 (top). The author with a typically coloured male of about 23 lb., lower Sustut River, September 30, 1986. This was the largest of approximately 140 steelhead caught by Ron Tetreau and the author between September 25 and 30 that year as part of a population estimation program.

Page 211 (bottom). Typical female steelhead of about 8 lb. caught and tagged near the outlet of Johanson Lake on September 30, 1986.

Page 212-213. The lower Sustut Valley as it appeared halfway between the Bear confluence and the Skeena in September 1991, before the loggers arrived to alter the landscape and viewscapes forever.

Page 216. The reigning world record steelhead of 42 lb. 2 oz. caught by David White near Bell Island, Alaska, on June 22, 1970. This picture was copied from the March 1971 edition of *Alaska/Life on the Last Frontier.*

Page 217. Chuck Ewart holding his 36 lb. Kispiox fish shortly after it was caught on October 5, 1954. Photograph provided by Northern Hardware in Prince George, British Columbia.

Page 218. Carl Mauser's 33 lb. fly-caught steelhead, Kispiox River, October 8, 1962. Photo taken from a copy of Mauser's personal diary given to the author.

Page 222 (left). The unidentified woman from Alberta with her 42 lb. Shames Bar steelhead caught on Labour Day weekend, 1984. Picture provided by the photographer, Gord Ronald of Prince Rupert.

Page 222 (right). The Andy Bouchard (red hat) 38.5 lb. fish taken at Esker Bar on September 13, 1986. Photograph provided by Andy Bouchard.

Page 222 (bottom). Daniel Daigle with the 48 in. steelhead caught near Esker Bar on the lower Skeena on September 17, 1993. Photograph provided by Daniel Daigle.

Page 223 (left). Robert Johnson, the long-serving test fishery contractor of the day with the 38 lb. steelhead taken on August 1, 1998. Photograph provided by DFO, Prince Rupert.

Page 223 (right). Contractor Bill Kristmanson with the 42 lb. steelhead on the test fishery wharf at Tyee on the lower Skeena River shortly after it was captured on August 1, 1998. Photograph provided by DFO, Prince Rupert.

Page 224. Pam Pitzman holding the fish caught by her husband Fred on October 11, 1976. Pam is 5 ft. 5½ in. tall. The same careful, proportional analysis performed on the Daigle fish pictured previously gave a conservative estimate of this fish's length of at least 48 in. Photo by Fred Pitzman.

Page 225. The author weighing a large Skeena fish with the equipment that put to rest all doubt about estimates derived from formulae based on length and girth measurements. Photo by Oregon Bob Hooton.

Page 234 (top). Colleagues (l to r) Dionys DeLeeuw (RIP), Mark Beere and Colin Spence along the Ain River, Haida Gwaii on April 8, 1989. The Misty Isles were once well populated with such spruce trees.

Page 234 (bottom). September 16, 1993. Ken Belford in conversation on the virtues of the Damdochax from the porch of the historic telegraph trail cabin at the base camp established by him and Alice on Damdochax Lake.

Page 236. In 25 years my son has seen some of the best river fisheries in British Columbia undergo a one-way transformation at an ever-accelerating rate. What will his legacy be?

AUTHOR BIOGRAPHY

Bob Hooton was born and raised in the Vancouver area and spent all his school years there, including attending Simon Fraser University as a charter student. Entry to a career in fisheries came a year after graduation. His career path took him from the East Kootenays to Victoria, then Nanaimo and on to Smithers before returning to Nanaimo in the later stages. Steelhead and steelhead management dominated his professional and leisure pursuits. A second university degree from the University of Idaho was a milestone of his first tour of duty in Nanaimo.

After 37 years of public service, Bob retired. Following that, he authored his first book on the history of management of the iconic Skeena River steelhead, *Skeena Steelhead – Unknown Past, Uncertain Future* (Frank Amato Publications, 2012). Much remained to be recorded about other important rivers in the province, however. In this second book he describes his personal association with some of them.

Bob and his wife Lori live in Nanaimo, British Columbia. They have three grown children.